W0114896

# Scripting for the New AV Technologies

## Second Edition

# Scripting for the New AV Technologies

## Second Edition

**Dwight V. Swain**
**Joye R. Swain**

Routledge
Taylor & Francis Group

NEW YORK AND LONDON

First published 1991 by Focal Press

Published 2023 by Routledge
605 Third Avenue, New York, NY 10017
4 Park Square, Milton Park, Abingdon, Oxon OX14 4RN

*Routledge is an imprint of the Taylor & Francis Group, an informa business*

Copyright © 1991 by Dwight V. Swain and Joye R. Swain.

All rights reserved. No part of this book may be reprinted or reproduced or utilised in any form or by any electronic, mechanical, or other means, now known or hereafter invented, including photocopying and recording, or in any information storage or retrieval system, without permission in writing from the publishers.

Notice:
Product or corporate names may be trademarks or registered trademarks, and are used only for identification and explanation without intent to infringe.

**Library of Congress Cataloging-in-Publication Data**

Swain, Dwight V.
    Scripting for the new AV technologies / by Dwight V. Swain and Joye R. Swain.—2nd ed.
      p.  cm.
    Rev. ed. of: Scripting for video and audiovisual media / Dwight V. Swain.
    Includes bibliographical references and index.
    ISBN 0–240–80071–0 (paperback)
    1. Audio-visual materials—Authorship.   I. Swain, Joye R.
II. Swain, Dwight V. Scripting for video and audiovisual media.  III. Title.
LB1043.4.S9   1991
371.3'35—dc20                         90–28850
                                       CIP

**British Library Cataloguing in Publication Data**

Swain, Dwight V.
    Scripting for the new AV technologies.–2nd. ed.
    1. Audiovisual materials. Scripts. Composition
    I. Title  II. Swain, Joye R.  III. Swain, Dwight V.
    Scripting for video and audiovisual media
    808.066

    ISBN 0–240–80071–0

ISBN 13: 978-0-240-80071-4 (pbk)

To Charles Nedwin Hockman . . .
my friend, who drove me crazy for forty years—and
taught me the business.

*D.V.S.*

To my mother, Opal Cowan Garland . . .
who taught me to read, to think rationally, and to love.

*J.R.S.*

# Contents

# Preface

Never before have there been so many opportunities in audiovisual. And never before, if you've prepared yourself properly, has it been so easy to get a start in the field.

Why? Because in the ten years since the first edition of this book was published, the world of audiovisual has exploded. Entirely new technologies have come into existence—from laser film to compact-disc interactive. As one expert comments, "Don't be afraid to try something because you don't have the equipment to do it. If you think it up, someone will figure out a way to get it to the viewer."

In consequence, audiovisual is everywhere—a $19 billion industry, not including broadcast TV or cable. (These add up to another $15 billion.) Tom Hope, a top gatherer of statistics in the field, estimates that there are some 8,600 production companies turning out programs and, as well as anyone can figure, more than 440,000 people are employed in the industry. There are at least 650 postproduction firms and 300 laboratories. Five thousand production units in 4,200 corporations produce AV materials, as do 1,500 government agencies, 1,000 colleges, and 50 religious organizations.

Hospitals use AV programs to orient patients and train doctors. Children from kindergarten through high school learn about everything from the alphabet to calculus, and graduating seniors watch persuasive university recruitment presentations. Airports feature random-access videos to inform arriving passengers about local attractions. National parks offer guidance and information via interactive video kiosks. The U.S. Army uses interactive computer-based programs and interactive video disc training materials to instruct soldiers on subjects ranging from equipment maintenance to battle tactics. The U.S. Postal Service has developed a variety of AV approaches to train personnel to operate and maintain the complex sorting machines that have speeded up the handling of an incredible volume of mail. Video walls are increasingly common in department stores and conference centers. Multi-image slide shows attract tourists or businesses to sponsoring communities. Computer games instruct and entertain all ages.

Even Santa Clauses are trained with videos!

What does all this mean to the scriptwriter?

More work.

More variety in the types of work available.

More interesting opportunities.

And a need for more knowledge of just what these fascinating new technologies have to offer.

That's what this book is all about. Because scripting for computer-generated programs, laser discs, digital video interactive, multi-image, video walls, and all the other new technologies—as well as the old standards like slide shows, filmstrips, and linear video—requires new techniques and handling on the scriptwriter's part. Today, you can even combine overhead optical projectors with computer-generated graphics thrown on the same screen or ceiling.

That all makes it a new AV world out there. It's exciting. And the number of opportunities are increasing every day.

Is there a place for freelancers as well as staff people? Indeed there is. The door's wide open.

So, welcome to the new AV world—and good luck!

# Acknowledgments

Too many people have helped with this book to name them all. But at least the following deserve special mention:

Thomas H. Bair, Kinton, Inc.; Nancy Chess, J.D. McCarty Center for Handicapped Children; *Computer Video Decisions*; Electrosonic Systems, Inc.; Linda Ellingboe, U.S. Postal Service; Jeanne Flanigan; Frank Garber, United States Video; Kevin J. Gillen, Capitol Video Communications, Inc.; Wesley H. Greene, International Film Bureau, Inc.; Jeff Griffin, Charles Machine Works, Inc.; Craig Guthery; Jack Hirschfield, U.S. Postal Service; David Hon; Dennis Holm, U.S. Postal Service; Thomas W. Hope, Hope Reports, Inc.; Glen Hopman, Smithsonian Institution; Warren K. Jordan, Jordan Associates; Thomas Kleiman, National Park Service, U.S. Department of the Interior; Jenifer Litwin, U.S. Information Agency; Layton Mabrey, U.S. Postal Service; Ann MacFarlane, U.S. Information Agency; Molly Maloney, Dayton Hudson Department Stores; Dave Mathis, U.S. Postal Service; Rockley L. Miller, *The Videodisc Monitor*; Garland McWatters, Metro-Tech; Lynn Allan Paschal; PDC Media Productions, Inc.; William E. Pryor; Jim Randle, U.S. Information Agency; Stacey Rose, U.S. Information Agency; Ariel Schwartz, Exhibit Technology; David Snyder, Marshall Fields; Sue Sorenson, Dayton Hudson Department Stores; Michael Hastings Spencer, University of Maryland; Kristin Staubitz, Dayton Hudson Department Stores; Rachel Stevenson, International Film Bureau, Inc.; Victor H. Stonick, Essex Corporation; Richard Thorp, Veterans Administration; Karen Tomlinson, AT&T; Benjamin S. Walker; Will Wright, Maxis Software.

To these and all the others—thanks!

# Scripting for the New AV Technologies

*Second Edition*

# PART ONE

# YOUR AV CAREER

# —1—
# You: Scriptwriter

What's a scriptwriter?

A scriptwriter is somebody who translates ideas into a form that can be photographed or recorded.

He or she is a conceptualizer, a developer, and an organizer of information and intentions as well as a writer.

What's an *AV* scriptwriter?

An AV scriptwriter is a member of the production team of an AV—audiovisual—project.

What does a scriptwriter do?

He or she writes a script to guide the production team and the actors. The script tells the production team what to shoot and what sounds to record and tells the actors what words to say and how to say them. It also indicates how the pieces should be joined together to form a whole.

Which is to say, a good script is the backbone of a good production.

But it's a long road from idea to finished product. And, actually, the process begins even before the idea. Thus, as a scriptwriter, frequently you're called in by a producer or a client—that is, a sponsor for a program—when the project is first considered. All you'll have to go on at this point is that somebody thinks that, for whatever reason, some sort of AV production would be desirable.

Your task then is to look at the potential client's needs and notions and to determine the purpose of the production. Why does the client want it made? What does he or she hope to accomplish by so doing?

Equally important, who does the client want to show it to? What change in thinking or behavior does the client hope to achieve in relation to this target audience?

Once you have these pinned down (and we'll talk a lot more specifically about them in Chapter 4, "The Issue of Audience"), you'll zero in on the program's length, the circumstances under which it will be viewed, budget limitations, and any special client requirements.

The list goes on and on: Which medium will or should be used—simple slide show, videotape, computer-based training, or whatever? Will lip-sync or narration be preferable? How about animation? Or would a game simulation prove more effective?

Ideally, when a scriptwriter is called in, all such organizational decisions on the project have already been made. Often, however, they haven't, so the writer is forced to take over and pull the project together.

Frequently, too, the writer will also be involved in all aspects of production, from determining specific program content to rewriting scenes so that they can be shot with cheaper sets.

Since a good deal of AV is aimed at instruction or providing information, it's desirable for the writer to have some familiarity with educational theories. Is it enough to "tell them what you're going to tell them, tell them, then tell them what you told them"? Or do you need to provide opportunities for the audience to respond and interact? What kinds of examples help people to learn and remember? Can a "talking head" teach as well as closeups of a motor, or is a diagram better? Will visual or auditory stimuli be retained best in this particular situation? Does humor help instruction, or is a serious presentation more effective?

Add all that to the job of being *creative*—that is, coming up with different or unusual approaches. When asked to devise a more effective way to present the history of science to junior high school students in 30 minutes, I pondered at length and finally came up with a tongue-in-cheek pseudo-folk song that provided background for a character who wandered through assorted historic periods with a light popping on in his head as curiosity led him to key scientific truths. In another program, designed to teach quick and accurate decision making under stress, trainees had to respond to the plight of the victim of a simulated accident; correct responses saw the victim getting better, incorrect, worse, just as in real life.

All of which makes the scriptwriter a vital member of the AV project team. Even though that role is wearing on occasion, it's also stimulating and challenging. Few experiences can match the excitement of the flash of insight that comes with finding a fresh approach to fire a motivational video or a new twist of computer-based training to teach math to sixth graders or the warmth of knowing your work has helped a family cope with a cancer diagnosis.

How do you get a start as an AV scriptwriter?

You took the first step when you opened this book. It will help you master the basic skills you need and grasp the pattern behind the field's technical complexities. Beyond that, the answer is practice and experience. What you need to do is immerse yourself in AV every chance you get. Seek out production crews and volunteer to help in any way you can. Talk to people in the business. Read journals and books on the subject. Get a job in the field—any job, even if it's only sweeping the studio floor.

What about the job-versus-freelancing issue?

In the first place, it's quite possible you'll have no choice in the matter.

Without experience, you have little to sell, but you *can* write scripts on your own in hopes of interesting someone. Or, you and a few friends may produce a program as your donation to the United Way or such. All these freelance things, in other words, will get you started. But they also may lead you to a steady job.

When the job comes, it may be in a hospital or university or corporate media unit. Then you may freelance evenings and weekends, selling scripts to outside clients.

Actually, relatively few producers handle the kind of volume that calls for a full-time writer on the payroll. More common is an arrangement by which a freelancer has a contract to supply scripts to a corporation or government agency for the length of a grant or project. For its duration the writer may even be on salary, reporting to work regularly as part of the staff but let go at the end of the operation.

Other scriptwriters freelance full time. That is, they work independently and contract with production companies to prepare scripts as needed. They're paid by the job and work at home or in their own offices as piecework contractors.

Each arrangement offers its own advantages and disadvantages. We'll discuss them in more detail in Part Four, "Succeeding as an AV Scriptwriter."

# 2

# Mapping the Field

You're ready to work in AV—but what exactly *is* AV? What does it include?

Let's take those questions one at a time. The answer to the first is easy: Audiovisual, in our context, refers to any recorded presentation that involves both audio (sound) and visual (pictures).

What it includes is more complicated. Consider any topic. *XYZ Dynamics,* let's say.

Now, add a reel of unexposed motion picture film. Or a reel of videotape. Or a videodisc.

How does information about the topic get on film or tape or disc? What's the process?

More specifically, who plans it?

The answer, in each case, is the scriptwriter—even though, on occasion, this person may go by other names.

What the scriptwriter does, in turn, depends on the technology by which the information—the topic—is set forth.

If the presentation is to be made on film, images are recorded on a chemically treated strip of celluloid. The scriptwriter (probably in collaboration with a subject specialist) describes the shots and the order in which they're to appear, plus the words and sounds that are to go with them. These words serve as a guide for the technical people assigned to the project—the director, camera operator, editor, and so on.

What about videotape?

Here the sounds and images are recorded electronically on a strip of magnetic tape. As in film, the scriptwriter describes what the audience is to see and hear. Following these instructions, the video crew takes the desired pictures and records the sound and arranges the shots in the proper order.

And videodiscs?

A videodisc is a thin, circular plastic plate. Images and sounds may be stored on it for playback on a monitor.

Film and tape are *linear* media. That is, the presentation goes in a straight line from beginning to end. To show a segment from the middle, say, the film or tape must be run forward or backward to the desired place. This takes time and limits the usefulness of the media.

A videodisc, in contrast, can respond quickly to electronic commands to move from one segment to another, so it lends itself to *interactive* presentations. It allows the viewer to change the order in which sequences are presented, picking and choosing between them and, if desired, leaving some out. Since the sequences may be of different lengths, this also may vary the length of the presentation.

In consequence, conception grows more complicated and the role of the scriptwriter changes. In film or videotape production, the scriptwriter by and large conceives and carries through development of the script from beginning to end, even though a variety of other people may have input.

In videodisc, a more complex planning process is followed. A major part of the process of conception is placed in the hands of the *instructional designer*.

This individual works in cooperation with another, the *content specialist*, who determines the specific content to use, designates examples, and such. The specialist decides what images are to appear on the screen and from which camera angle, how long they are to remain, the order in which the sequences are to be presented, and the like.

To further complicate the matter, much of the work in making a videodisc presentation is provided by a computer, an electronic brain designed to store and sort out information. A *computer programmer*, a specialist who prepares a program *(software)* to develop an organizational pattern for retrieving selected data, is in charge of this aspect. Using preprogrammed software to prepare a program, whether training, educational, or entertainment, is called *authoring*.

In consequence of all this complexity, scriptwriters frequently find themselves limited to describing action or settings and writing down the words to be spoken.

This is not always the case, however. In small operations especially, scriptwriters may work to the limits of their initiative, originality, and ability. Quite possibly they'll end up as instructional designers themselves!

So much for the overview. Now, specifics. As a scriptwriter in audiovisual, you may work on the following:

*Slide shows.* For the purposes of this book, a simple slide show is a presentation of still photos in a predetermined order and length of time on screen. Thus, filmstrips really are slide shows. Sound may be prerecorded or spoken by a narrator with or without a script.

*Multi-image displays.* A presentation where the viewing area contains more than one image at the same time. Most often, the designation is applied to slide shows using more than one projector. However, the same effect can also be attained with split-screen video or film, video walls, or even computer monitors.

*Video walls.* A variation on multi-image shows, video walls split a single video signal so that a programmed, enlarged series of still or motion pictures are projected on a gridded series of TV monitors.

*Linear video/film shows.* A presentation designed to be played from beginning to end, in the manner of the familiar "movie."

*Multimedia shows.* Any presentation that combines two or more media. They include slide shows that provide a background for live actors on stage; the many museum displays that mix slide shows, video, and exhibits; and viewer participation spectacles such as the *Earthquake* ride at Universal Studios or *The Great Movie Ride* at Walt Disney World.

*Interactive video/disc programs.* Shows that permit the controller or viewer to change the order in which sequences are presented or even to eliminate some sequences.

*Computer-based training/games (CBT).* Interactive presentations in which branch-activating prompts allow the viewer or operator to move through or select parts of a training program.

Each of these types of presentation has its strengths and weaknesses, its good and bad points. Deciding which one to use on a given project can be a difficult matter. As a general rule, however, the more complicated the presentation, the more expensive it becomes.

The slide show has the virtue of economy. It's easy to produce and to use, and it handles simple messages very well. Changes are inexpensive. But it tends to have an out-of-date flavor, it lacks impact, and audiences find it less than spectacular.

Multi-image shows come on stronger, but may be difficult to transport and handle if more than one projector is used. Equipment glitches are an ever-present danger.

Video walls emphasize color and spectacle. They're favored for conventions, trade shows, and the like. But they're complex and expensive, and they demand highly skilled technicians to put them together. They're better for making a flash than for presenting detailed information.

Linear video/film shows can be about as simple or as complex as the client's budget will allow. Video is markedly cheaper under the right circumstances, but film provides a sharper image. The decision is based on what's important.

Like multi-image presentations, multimedia shows can prove costly, especially if they involve combining live actors with audiovisual surroundings. But you can't beat them for impact and attention-holding.

Interactive video/disc programs can handle virtually anything you can imagine. Their cost can be well-nigh astronomical, however, and it's difficult to update them, because once anything's on a disc it's pretty much permanent. The lack of equipment standards poses another problem. A program designed for one company's equipment often can't be played on a machine designed by another company.

Computer-based training is highly effective. Advances in transferring video images to computers and other technologies make this a field with an incredibly rapid rate of growth. And although programmers don't come cheap, one person using an authoring system can produce a very complex, impressive program at home with no more equipment than a computer.

Obviously a field as broad as audiovisual is complex and wide ranging. Yet its basic principles are relatively simple and well within the grasp of any intelligent person willing to work conscientiously to learn its specialized vocabulary, concepts, and procedures.

# 3

# The Project Team

A lot of people are involved in any major audiovisual project. As a writer on even the simplest program, you'll ordinarily find yourself working with

the client or sponsor,

the subject specialist,

the producer,

the director.

There'll also be a technical crew, of course—camera people, lighting people, sound people, editors, and so on. On more complex assignments—interactive video, for example, or computer-based training or many multimedia extravaganzas—you'll encounter at least two others: the instructional designer and the computer programmer. Let's look at these one at a time.

*Clients and sponsors* come in all shapes, sizes, and dispositions. Some are a dream to work with; others, anything but. I'll explain my reasons for saying this a bit later in this chapter. Meanwhile, we'll simply define the client as the person or persons who want a program produced.

The *subject specialist*, sometimes called a *content specialist* or *technical advisor*, is someone who allegedly knows all about the content of the program. This is the one who tells you (or is supposed to tell you) anything you need to know about the material to be included in the program—the person to whom you turn for information when you're stuck. It's the specialist's assignment, too, to steer you clear of incorporating embarrassing errors of fact or interpretation in the program and to keep you posted on client attitudes, policies, and the like. The specialist also plays a major role in saving you from misusing the jargon of the trade.

The *producer* is the man or woman in charge of seeing that the program is carried through to completion. Producers ordinarily are familiar with all phases of the operation, but not always. Frequently, the budget is their primary concern. They tend to think in dollar signs and to reject any scene or shot that may result in a cost overrun, no matter how aesthetically vital it may be.

The *director*, of course, directs. In effect, the director is the gang boss of the production crew, and decides what pictures should be shot and from what angle. If live actors are to be included, the director is the one who tells them what to do, how to do it, and how to say any words you put into their mouths.

Will the director follow your script? Not necessarily; indeed, not even probably. Directors certainly will apply their own interpretation to a script. A saying in the trade has it that the script doesn't really exist until the director has his or her way with it. This may be good, and it may be bad. Many times, I've had scripts improved by a thoughtful director. On the other hand, a careless or less than competent director may totally destroy you, simply because he or she hasn't bothered to read or think through your work. But more of that later. Meanwhile, just know that often there's very little you can do about the director's changes, save perhaps to clench your teeth and pray a lot.

The *instructional designer*, as mentioned in Chapter 2, is a specialist you'll encounter mostly in building interactive programs; the designer's influence can hardly be overestimated. Experts on manipulating words and pictures to modify viewer attitudes and behavior, their goal is to set up subject material in such a pattern as to shape audience thinking in a desired direction. In consequence, they will either work with you on your script, laying out the shots and lines through which it's to be developed, or they will set up specifications for you as to how it's to be put together.

I should also point out that the instructional designer's influence is spreading. Although currently working primarily in the interactive field, don't be surprised if the instructional designer also turns up in anything from slide shows to linear video.

The *computer programmer*, another expert, will design any software to be incorporated in the program. This is a vital role where interactive programs, multimedia shows, video walls, and even programmed slide shows are concerned. You'll need to know what the system you're working with can do, but do leave the programming itself to experts! "Authoring" systems are available that allow computer-literate nonprogrammers to prepare software, but their scope is limited. A real programmer can provide much more variety.

What about editors, camera people, lighting experts, set designers, actors, and so on? Your contact with them will be limited, in all likelihood. But you do touch base with them, so just bear in mind that their work, too, is important and that they deserve respect and courtesy.

*   *   *

Now, take everything I've said here and stick the words *maybe* and *sometimes* in front of it, for the degree of interplay on any set is great. For example, the director is supposed to tell the cameraperson what to shoot, but quite possibly the cameraperson may have ideas of his or her own. If the director is weak or uncertain, the cameraperson may end up making shooting decisions, for better or worse. Or a superior sound person may dominate line delivery; or a topnotch lighting person, the mood to be created. In other words, this is a field in which few absolutes exist, and every job is handled a little differently than every other. It's a thing to bear in mind before you decide what's right or wrong in a given situation.

Further, all these people are important to you, believe me—not only in your present work, but also in getting future work and lining up jobs. If, for example, you've made a good impression via friendliness, cooperation, or knowledgeability on a potential client or sponsor, that person will quite possibly be inclined to think of you when looking for a scriptwriter in the future. And double that for advertising agency and public relations men and women.

This is also true for producers. If you remind them of your existence by dropping by or sending an advertising postcard every month or so, sooner or later one may decide to give you a shot at a project.

Similarly, an instructional designer or computer programmer or content specialist or director may suggest your name in the course of talking over a job.

This is equally true of every member of the crew. People who like you are likely to give you a call when the grapevine reports a new film or video or disc on the horizon, long before any public announcement is made. I can recall a number of assignments that came my way because people, out of the kindness of their hearts, let me know that a grant had come through or a contract had been signed.

Understand, however, that this kind of networking must of necessity rest on the solid bedrock of your own competence. People will not go out of their way to help you if you can't deliver. And once the word gets out that you're a talker, not a doer, times will be rough indeed for you.

At this point a few practical observations and experiences may prove enlightening and help you to cope with the other members of the project team. Consider, for example, a hospital picture I worked on with a very talented director. Everything went beautifully until he called me in to look at the rough cut.

This was supposed to have been a public relations film for the hospital, but what I saw on the screen was a twentieth century chamber of horrors. Hogarth's portrayal of Bedlam in A Rake's Progress will give you the general idea.

I screamed. Loudly. The director made appropriate excuses, apologies, and placatory gestures. We went back to work, scrabbling up pieces of film from the cutting room floor this time. Anything that showed a smile or sunlight found a place. Eventually, the picture was completed and accepted.

Years after the project had ended, I found myself sitting down for a drink with the director's former wife. The conversation turned to the near disastrous hospital job. I found myself expressing my bafflement at what had happened in the rough cut.

The ex-wife downed half her drink in one gulp. "You know," she said, "I almost talked to you about that at the time. But Eddy"—not his name—"and I were still married then, and I felt it would be disloyal. What you didn't know was that Eddy was walking a tightrope those days. His health—well, he kept thinking he might have cancer. So what he shot were his own fears, even though he probably didn't realize it."

The lesson here is that, as a writer, you work with people. All of these people have their own quirks and foibles, and you need to be aware of this before you touch pencil to paper.

This being the case, you need to ask yourself certain vital questions.

## WHO'S PAYING?

You've heard the old line about "He who pays the piper calls the tune." Double that for the person who lays out cash for a script, whether it's the client/sponsor or the producer. Here we'll be speaking primarily of the client/sponsor.

Too frequently, the client believes that the pen that signs the check is also flowing with literary genius. More often than not, said client won't hesitate to explain your script's shortcomings to you and to suggest items to be inserted or excised.

Now, criticism is fine. We all can use it. But an AV scriptwriter is supposed to be a sort of expert—a master (or at least a journeyman) where a particular field of communication is concerned. It's assumed scriptwriters do what they do for a reason.

It follows that when, against your advice, Moneybags decides to costume machine parts as brownies or to delegate the narration to a harelipped relative, it may have some effect on the total impact of the package.

What can a writer do about this? Often very little, I regret to say, unless you count biting the bullet as constructive action. "Take the money and run" is a saying that works to cover this contingency. It means, essentially, do what the sponsor wants, no matter how stupid, collect your pay, and go on to another project.

Actually, your sharpest weapon may be the fact that Moneybags has money and he or she probably didn't get it by being a fool. So if you point out that the client stands to lose money or minimize profits on a sizable investment, quite possibly he or she will see things your way. This is especially true if your arguments are logical and well presented. If you protest gently and politely, you might save the job and be hired for another by the same client.

You also may find it helpful to listen with an open mind to what your patrons have to say. Assuming that they're knowledgeable regarding your subject, it's possible they can make a useful contribution.

It's worth noting, too, that many individuals who finance AV programs have only the most perfunctory interest in scripting or production. Where they're concerned, delegating authority is the rule. And that brings us to another question.

## WHO'S GOT THE MUSCLE?

Again, we're going to be talking first and foremost of the client. But a producer or director may also fill the bill, especially where the matter of ego comes under consideration.

Like it or not, script decisions must be made for any AV project. The question is, who's going to make them? And, having made them, will he or she stick by them?

These are terribly important questions. Because if the wrong person—that is, one not authorized—lays down the law to you, or if the right person proves too weak or easily influenced, trouble may be your workmate for the rest of the job.

What specific types are we talking about? We can divide them into three groups: (1) the power-hungries, (2) the indecisives, and (3) the impossibles.

The *power-hungries* are people like the vice president of a firm for which I did a consulting job. He'd gained his rank by marrying the daughter of the chairman of the board. Now he wanted to prove his merit.

The company president, in turn, took a dim view of the situation. So far as he was concerned, the vice president was an upstart and a phoney.

My project became a bone of contention between them. Each wanted to run the show, determine its content and presentation. Anything one would approve, the other would veto.

The completed job was a mess, of course. It could have had no other outcome.

This isn't to say that it can't also be a good thing when someone seizes power. I've seen several floundering projects saved by some individual who, taking command, insisted that all work meet his or her approval.

The second group, the *indecisives*, present a different problem. They're the folks who can't make up their minds. Most often, the reason is that they're afraid of being wrong. You find them in droves in government and occasionally in major corporations: living embodiments of the old saw that responsibility's assigned, authority's seized. Presented with a script, they'll neither accept it nor reject it; they'll only pass it on to someone else to make the decision.

The *impossibles?* You might define them as individuals who lack the strength or drive to seize power, yet still are determined to have their way with those beneath them—that is to say, the innocent scriptwriter who strays into their web. Demanding endless changes for change's sake, they make it impossible to complete the program satisfactorily. Nor do they particularly care. Ego inflation is really all they're after. My own pet example of the genus is an academician of great prestige who had a large foundation grant to finance a series of presentations. At the end of each meeting, I was sent forth to work up a concept. By the time I returned for the next appointment, however, the academician had come up with a new and allegedly more brilliant approach. He could not understand why I found this both discomfiting and (since time is money) financially disastrous. Three times and out. I withdrew from the project, as did the next couple of writers he hired. I don't know if the job ever was completed.

Just in case you feel, after this recital, that I'm too negative, let me throw in a few stories with happier resolutions.

One involved an aide to the president of a national trade association for which I was assigned to script a major project. The president was pompous; the aide, anything but. Every time the president would start to throw barriers into my path, the aide would skillfully skate around them. I never had a job go with less friction. When all was over, the president was convinced he'd been in charge, but the aide and I knew better.

At a large company, the media branch—which is to say, the AV section—became embroiled in a row with the public relations department. A clever executive's solution was to hire me to "script" the project over which the fight had arisen, even though this scripting involved little more than changing commas. But my name drew screen credit, so each contending side could, by devious reasoning, claim it had won the victory.

What all this adds up to is that there'll be decisions to be made on any project. You, as writer, should stand aware of this and insist that some one individual—the technical advisor and subject matter specialists are good bets, since you'll be working closely with them—have the power to call the shots.

This will have two beneficial effects. First, it's remotely possible (though certainly don't count on it) that the sponsor will stick by his or her word and *let* the individual decide. Second, it should at least reduce the number of people with whom you have to deal in the preliminary stages.

## WHO'S DOING THE JOB?

Remember the director I described at the opening of this section? He has a lot of relatives.

Let's divide difficult directors into four categories: the artistes, the egomaniacs, the incompetents, and the procrastinators.

What can one say about *artistes?*

Well, for one thing, they're artistic. But their definition of their talents might differ a bit from yours or mine. Thus, the view of "dramatic irony" taken by one man on a feature film I scripted almost brought us to blows. He insisted that the plane bearing the two young lovers into the sunset at the conclusion crash in flames.

Is there an answer to this kind of thing? Not really. Certainly it doesn't warrant barring such people from the trade. The people who are the most difficult to work with frequently come up with extremely clever approaches, which a more practical mind might never have considered. The best advice, probably, is for you to grip your own sanity tightly, then take full advantage of any coruscating concepts the artistes may have to offer, while at the same time calming their more egregious excesses.

The *egomaniac* rates as much the same as the artiste, save that he or she has no artistic pretensions. Egomaniacs just want to do it their way— on whim and without forethought—and of course it's right because it's theirs.

The catch with this is that the sponsor rates a little consideration too. You've been paid to come up with an idea, a concept, an objective, and a script. You've thought them through, worked them out, honed them down till they mesh like watch gears. For the producer or the director or anyone else to throw all that out on the chance that his or her own slapdash notion will satisfy the sponsor better is neither intelligent nor fair.

What do you do about it? One nasty gambit I've come up with is this: I insist that I, the writer, have a separate contract—one that states explicitly that I'm in no way responsible for the finished project. Of course, you can only do this when you're hired as a freelancer. If you're working for a production company, you simply have to go along.

One result of this tactic is that failure to follow the approved script thus becomes the producer's headache, not mine. Indeed, the producer may be hesitant to deviate from the script in any way without approval from on high, which of course doesn't necessarily mean me, but this at least leads to an exercise of reasonable restraint. In addition, sponsors seem to like the idea. If things go wrong, it tells them who's to blame.

Much the same situation obtains where the occasional incompetent producer or director is concerned. One that I knew apparently was afraid of closeups. He simply would not fill the frame with any image, no matter how clearly it was required. As a result, key shots lost impact, and a potentially strong filmstrip turned to milk and water.

In another case, a former portrait photographer lighted faces beautifully. But his shots all came through as what sometimes are termed *talking heads*. He couldn't seem to compose a visual to make a point or to frame an idea.

Then there are the *procrastinators:* the workers who, otherwise competent, simply put off and put off and put off.

When pressure eventually erupts, these dawdlers' scruples too often fly out the window. In at least one case, I lost a valued client because a producer, whom I thought was my friend, in my absence glibly blamed delays on me.

Again, the solution in most cases is to stand firm for separate contracts if you can. Don't make your pay contingent upon the project being completed.

You also should think twice before you agree to front for a producer you don't know. Just because someone telephones to tell you they'd like you to work with them on a project doesn't mean that you necessarily want your reputation linked with theirs.

I probably should conclude this section with an across-the-board apology to all the fine producers and directors with whom I've worked: *Truly, friends, I've no intention of slandering you. You're good, you're solid, you're conscientious, you're creative. None of my remarks in regard to artistes, egomaniacs, incompetents, or procrastinators is aimed at you. And I fully recognize that just as no producer or director could get very far without a script, no script would mean much without you to translate it into sounds and visuals.*

The thing is, writing is work—the writer's work. To survive, writers must face the dark side as well as the bright, and they must treat writing as a business. We'll discuss this in more detail in Chapter 15, under "Deals and How to Make Them."

## TALKING TO TECHNICAL ADVISORS

I wish I had a dollar for every time my neck's been saved by a technical advisor. Whether you call them *subject matter specialists, information experts, content specialists,* or *project consultants,* they remain pearls beyond price where writers are concerned.

A technical advisor keeps you on the straight and narrow track in relation to your script's subject matter. If you're writing about Black Angus cattle, the advisor can tell you off the top of his head that they're natural polls. If the topic is electrical symbols, the advisor knows which sign designates a power source and which a resistor. Latin subjunctives? Have no fear, your expert's here!

The key fact to bear in mind is that the specialist's work with you quite possibly will take a lot of time. Both your advisor and the person who assigns him or her needs to accept this on the deepest possible level. A boss who thinks a chief aide can be both on the job and helping you is almost sure to end up irked if not bellicose. An advisor who doesn't realize you're going to preempt the hours he or she intended to spend working on the budget may not accept it cheerfully. And you, if you expect succor and don't get it, aren't likely to end up in the best frame of mind either.

I can't tell you how many times I've been signed up to a rush job only to find that the technical advisor's vacation started that day and he or she would be out of town for two weeks. So, be sure you have a competent technical advisor who will be available to you when needed.

While you're at it, it wouldn't hurt to feel out how your personality meshes with the advisor's. Sourpusses are seldom fun. Neither are strong, silent types who communicate in monosyllables and stand convinced that only a dolt could ask the questions you do.

This brings up the matter of how to go about asking appropriate questions. In general, your problems here will center on the individual who talks too little and the one who talks too much.

Where the person who talks too little is concerned, don't overlook a polite reticence. Early on, I strive to impress my advisors that I won't be embarrassed if they catch me in an error. In fact, I try to impress upon them that it's their duty to keep us both from being publicly humiliated by leaving such errors in until the final product (where they are terribly expensive to correct and become common knowledge). You'd be surprised how often advisors will not mention an obvious error because they don't want to seem rude.

A comment I often hear is a very hesitant, "Well, I guess that's right, but we'd never say it that way." Of course you want your narration or dialogue to sound natural to the viewers. You want to use their language. The only way you can do that, if you're not an active member of their group, is for your advisor to tell you how to say it. But first your advisor must be willing to tell you. That's not always easy.

Tape recording interviews will give you the vocabulary. I remember a project for a meteorologist. After two weeks, he approached me in amazement. "I didn't know you had a background in meteorology. How long have you studied it?" he asked. "Two weeks," I replied. But the real secret was that I'd recorded his answers to my questions and incorporated his vocabulary in my presentation.

The section on research in Chapter 7 offers another sound approach. What you need are questions to ask—the right questions. The "right" questions are those that reveal a certain amount of intelligence and background where your topic's concerned, while at the same time dredging up the data you need.

The reason you strive for this combination is that most experts are more prone to help someone who shows the potential of benefiting by their help than they are the individual apparently incapable of grasping any information they may convey.

At the same time, this doesn't mean that you must be an expert in order to ask questions. A reasonable command of basics—obtainable from rudimentary print research—is enough. And you certainly shouldn't hold back from poking into those aspects of your subject on which you need enlightenment. Indeed, as I've pointed out elsewhere (*Techniques of the Selling Writer*, pp. 277–8), your greatest asset in this area can be a willingness to be thought a blithering idiot. Ego, an unwillingness to let your inadequacies stand revealed in the cold light of day, can cut off your flow of data.

The other side of the coin, the person who talks too much, can prove equally as frustrating. Bubbling over with the subject, hung up on trivia, this type of expert may babble on till you're ready to scream.

A combination of flattery and advance planning may take shape as your salvation: "Man, you *really* know flipenroods, don't you? Let's get back to them after you've cleared up the hassenpfeffer principle for me."

The idea is that you *must* retain some semblance of control if you don't want a week's work to drag to a year.

On the other hand, I've known scriptwriters to simply record a long interview covering the process or instruction they were to handle, edit it slightly, add the shots, and turn it in. No one can say that their narration or dialogue doesn't ring true.

While we're talking about technical advisors and such, it might be wise to point out that not all necessarily will be subject specialists in the usual sense of the term. Some of the most useful—indeed, essential—technical advisors will come from the ranks of your production crew: your visuals people.

A good example may be found in the oscilloscope project described in Chapter 7. One of my wife's biggest problems on it was the fact that a vital element in the AV program was a "trace"—a point of light (often seen as a line) that moves across the oscilloscope's screen. To film said trace involved coordinating the camera speed with the trace speed. If synchronization weren't precise, the trace speed might be distorted or it even might not appear on the screen at all.

(You've seen a less complex version of this kind of thing in western movies when the wheels of the stagecoach appear to move backward because of the speed differential between camera and wheels.)

This created a writing problem. Could my wife describe the oscilloscope trace as behaving in a particular way and anticipate that it would so appear on the screen in the program? Or must she somehow write around it— and, if so, how could the program instruct students accurately and effectively in oscilloscope operation? Diagrams were a possibility, but she wanted to use the real thing if she could.

The solution came from a technical advisor—not from the oscilloscope side, but from production. His combination of mechanical and mathematical skills enabled him to work out a functionally effective procedure for filming.

## MIRROR, MIRROR

All through this chapter, we've focused on how you can best cope with your colleagues. Now the question arises: Is it possible your colleagues may have a little difficulty sometimes coping with you?

Let me cite three incidents that may illustrate my point. On a job in Washington, D.C., another scriptwriter was under discussion. "It's too bad," the producer said with real regret. "The guy's a good worker and he's got talent, but I can't afford to use him."

"How come?"

"He's simply not tuned in," the producer answered. "I mean, appearances can be important in this kind of town, and he's still stuck in a college-protest phase complete with sweatshirt and tennis shoes. I just can't afford to take him into a conference with men and women who take respect, neatness, and good appearance for granted. No matter how much ability he has, they'd automatically discount anything he said on the basis of his image. He wants to come on as creative, but he just looks sloppy. He wants to sound 'true to his art,' but what he sounds like is hard to work with."

I wish that were the only time I've run into this kind of thing, but it isn't. All too often beginners strike a wrong note just by their manner and looks. And because they make the wrong impression, opportunity after opportunity passes them by.

The second incident took place in Los Angeles, where I was discussing a possible script job with two local producers. They were hesitating. Obviously something wasn't quite right. Finally, after a cryptic exchange of glances, one asked, "How do you feel about style?"

I groped. "Well . . . I guess I never thought too much about it."

As one, the producers heaved a sigh of relief. "Thank God!" one said. "The last writer we had screamed bloody murder about us ruining his style every time we changed a comma." I still recall thinking how idiotic that writer must have been.

The third incident is perhaps even more to the point. The scene was Dallas, a government agency's AV planning conference. Out of a clear blue sky, an executive cracked a joke.

Everyone laughed—except me. For, as I realized with a start, I was the reason for—not the subject of—the humor.

Why? Because somewhere along the line I'd lost my sense of proportion. My voice had risen angrily. Whereupon, the executive had deftly poured oil on the troubled waters with her wisecrack.

I wish I could claim that was the only time when I've been difficult. Unfortunately, it wouldn't be true. But at least it doesn't happen as often now as it did when I was younger. The reason, I hope, is that I've begun to see such situations a little straighter and I've worked out more satisfactory ways of resolving them.

In the first place, I now recognize that AV scripting is, by and large, a cooperative venture. People other than the writer must, of necessity, get into the act.

I've also come to see that the world won't end if I don't have my own way on everything and not everyone will necessarily agree with even the most brilliant of script ideas. Indeed, no mode of attack is the "only" way to tackle a project. It's important that the writer realize this.

There's also the matter of your script itself. Have you worked it out with all possible care? Or have you conceivably written too fast or skimped on the research or gone a trifle slack on planning? Writers have upon occasion, you know, and you're only human.

In fact, being human, is it conceivable this isn't your best day? A cold, a headache, a disappointment—mightn't one or more of them have put you in a bad mood and rendered you difficult to deal with?

These are only questions, you understand, not statements. But they might be worth at least a moment of your consideration.

# PART TWO

# THE SCRIPT: PARTS AND PROCESS

---4---

# The Issue of
# Audience

Given a topic, an idea, or a message to convey, what's your audience and your purpose in relation to it? And why not talk about idea or message here instead of viewers?

The vital yet often ignored answer is that as a writer, you need to know who's going to see your program before you can meaningfully deal with the other factors. For, while you may play around with an idea or a concept or a message, the audience and what your client wants that audience to do after they see it is going to shape everything else if you're shooting for a truly successful presentation.

Take the topic "cancer." Your handling will be very different if your presentation is aimed at surgeons, patients, patients' families, researchers, or potential donors to a new cancer wing of the local hospital. "Drug use/ abuse" is a topic. What do you want to say about it and to whom are you saying it? To create an effective script, you need to know if you're addressing drug users or workers in a drug treatment center. Family members will require a still different approach, and so will police officers.

This means that, in attacking any topic, your first question will need to be, "What's the audience?"

*Audience*, as I use the word here, is another term for *special interest group:* the particular collection of people toward whom a program is slanted.

And yes, every program is—or should be—so slanted. Nothing is more surely doomed to failure than the presentation that presumes to appeal to the whole world. Even entertainment films try to call their shots in terms of specific groups: horror fans, bikers, yuppies, families, young people, action buffs, and so on. Only a handful of blockbusters succeed in casting their shadows over true mass audiences; in most instances, these box office hits are advertising and promotion achievements rather than cinematic colossi.

Projecting this principle into that segment of the audiovisual field with which we're dealing here, it may safely be said that a script is far more likely to succeed if you abandon the "general public" approach and write for a specific audience.

This is by no means as simple as it may seem. Ideally, the client will have an audience in mind: "I'd like a 15-minute multi-image show to rev up our salesmen to push the new Gizmo Eight." Most likely, however, you'll hear, "I want a really swinging thing about our company. We'll show it at sales meetings and to prospective customers and on television as a documentary and to potential investors in our stock. Oh, yes. Maybe we can show it at schools, too."

Often, it can truthfully be said that half the work in writing a script lies in finding out what the customer wants. That means establishing to whom the message should be directed. You can usually translate that, "Who does the client want to influence, and what does the client want the viewer to do after the presentation?"

Occasionally, if you're lucky, and especially in more expensive and complex productions, someone—the producer, the client's marketing department, or an advertising agency—will have done some sort of audience survey to determine potential viewers and their attitudes. Such studies may break down the target group as to ethnic background, financial status, educational level, you name it. They may zero in on people who live in a particular area, who subscribe to travel magazines, who own Fords, who have had heart attacks, or who are under twenty or over sixty.

Peripheral audiences may also be involved. I remember, for example, a slide show on which I worked many years ago that was designed to herald the glories of a college's new residence halls. Those of us who were involved sat through the approval showing. When the lights came on, the dean of students, who was in charge of the project, turned to me. "Good job. But there's one shot that's got to go."

He stepped up to the screen and pointed. "See?" The four young men and women seated at a table playing bridge remained totally innocuous in my eyes. "It's the cards," the Dean explained. "This show's going to be viewed by parents, not just kids. Small-town parents, lots of them, and this is the Bible Belt. They'll see those playing cards as the devil's picture book, with their sons and daughters on the road to ruin. So that slide comes out."

For more than forty years now, even though times have changed drastically, that scene's stuck in my mind. It helped to teach me that what you think of as your audience may not necessarily be.

Change of time and locale: to an early seventies narcotics conference in Los Angeles's San Fernando Valley and a film designed to turn young people off drugs. Again, the lights came on. A probation officer spoke.

"No good," he said. "This stuff's slanted to adults, not juveniles. It's phoney. That's why we called you. The kids know junkies and potheads. They'll laugh it off the screen. What we need is something that will speak to the kids."

These examples are obvious, of course, in order to help me make my point. But the same problem exists on subtler planes. When someone says he'd like an educational AV program, what audience does he want to reach? Students? Teachers? Parents? Administrators? Legislators?

Or, what about the travel show? Is it aimed at anthropologists . . . armchair adventurers . . . grant givers . . . or successful female executives with cruise money to spend? Is the State Fair display intended for passersby, or the boss? The in-house TV show, for foremen . . . line workers . . . management . . . public relations?

You can see how important this issue is. If you're fortunate enough to have a client who's zeroed in precisely on the target audience or you're working with a producer or other team member who's already settled this issue before you come on the scene, great. If not, before you start developing ideas, nail your audience down.

But it's not enough to know who your audience is. You also need to know their interests and prejudices so that you can devise an appeal that will yield the results your sponsor seeks.

Beware, too, of blind spots in your thinking. All of us have them. They're the attitudes formed years before—often in childhood—that we take for granted and that, too many times, we don't even recognize, but that others find offensive. Sexism, the tendency to downgrade women, is one currently in the spotlight and the focus of much "consciousness raising" in the feminist world. Racism is another issue that raises hackles, and so is agism, the assumption that senility or, at the very least, losing touch with contemporary reality automatically comes with advancing years. Derogatory ethnic or national humor and religious slurs reflect other blind spots. Denigration of mental or physical incapacity constitutes another segment of such lack of insight. And stereotyped responses to such inflammatory topics as abortion, drugs, or child pornography may blur clear thinking on the subjects.

Why do I call attention to these matters? Because they tend to elicit audience emotional responses so strong that they block out judgment or learning where your program is concerned.

It's important, therefore, that you become aware of such blind spots not only in yourself, but in your client or producer. All of you must recognize that words and modes of presentation have connotations as well as denotations and that ignoring them may negate your whole program.

Thus, an audience made of families with children may respond differently to a presentation that centers around infants than may one of swinging singles who feel that babies aren't worth the bother. Farmers don't react to rain the same way that urban residents do. Rich and poor respond to rent subsidies in separate ways. The scene a male audience finds hilarious or sexually stimulating may prove offensive to liberated women.

What do you do about this? The important thing is to check and recheck the attitudes reflected in your program, on both a conscious and unconscious level, so that you don't inadvertently slip in attitudes that may offend or distract your audience.

How do you identify your audience and determine its characteristics, its interests, its prejudices? What's the precise appeal that will yield the results your sponsor seeks? What makes such questions complicated is that, despite a fantastic amount of time and thought expended, no one has as yet perfected a formula that will give you the right answers every time. Indeed, there's not even agreement as to the best procedure to follow. Some experts swear by scholarly studies, surveys, polls, demographics, statistical analyses, and the like. Others ridicule these methods on the grounds that they're hopelessly academic and can never give proper weight to the many variables that are bound to crop up.

A public relations friend of mine with a reputation second to none preferred what he called the "hat trick." In doubt as to audience attitudes, he'd put on his hat and head out into the marketplace (that is to say, that area in which his target audience was prone to congregate) and drift about, chatting and asking questions. Most of the time, he got excellent results. But there were occasions when this sample proved too narrow and he missed the boat totally.

Another approach spotlights intuition—that is to say, you fly by the seat of your pants. Sometimes it works and sometimes it falls flat on its face. When my friend Fred Pohl launched a campaign to sell an apparently unsalable book with the obviously ridiculous copy capsule

HAVE YOU GOT A BIG BOOKCASE?
Because if you have, we have a BIG BOOK for you . . .

it sold out the edition—a clear example of the power of pure intuition.

For my part, I recall a major and expensive effort made at the behest of a sponsor to determine the attitude held by junior high school boys and girls toward science and scientists. The study was set up with a top sociologist to prepare questionnaires and supervise a nationwide survey. The results indicated that most of the youngsters considered scientists to be bearded old men in white coats who had no social life and were never home because they spent all their time trying to devise ways to blow up the world.

Fair enough. But I'm not at all sure that an afternoon or two spent talking to teachers or drinking cola in student hangouts wouldn't have given just as valid a picture.

There's no doubt, however, that the better you understand your target audience, the more likely it is that your AV presentation will accomplish your purpose. To that end, it's well-nigh vital that you should make it your business to become at least passingly familiar with the research educators have done on how people learn, in terms of such elements as perception, memory, concepts, and attitude change.

## ENTER LEARNING THEORY

Any new development is bound to breed speculation. Where audiovisual is concerned, this took the form, first, of theorizing and, later, research. This was especially true among educators, for they could see the new field's implications where teaching was concerned. Ideas and concepts proliferated.

There's no need to go into all the gropings and false starts this led to. In line with the old saying that no experiment fails—even an effort that doesn't work cuts off one more blind alley—the theorists moved ahead, increasing their grasp of the subject. Bruner, Soulier, Fleming, and a hundred others became respected names in the area. Research came to substitute for rule-of-thumb where audience attitudes were concerned.

Audience and needs analysis became the order of the day. Comprehension, perception, memory, and attitude were analyzed. Symbols and interpretations were scrutinized. Assumptions about learners and learning processes underwent reexamination. Programs were appraised in terms of assessment, diagnosis, and matching needs to instructional approach. Media effectiveness and the role of page layout were analyzed. The value of humor as a teaching tool not only was debated but studied scientifically.

In consequence of all this, a new specialist came into being: the *instructional designer*, the individual who determined the approach to be taken in a given program and the manner in which it was presented, frame by frame.

Further, the work of this specialist wasn't limited to the halls of academe. Applying behavioral science and learning theory to the audiovisual field, the instructional designer was accepted as essential to many corporate and governmental projects, an important addition to the project team.

Even the scanning of a single book like Fleming and Levie's *Instructional Message Design* will add to your awareness of how people learn.

## ENTERTAINMENT VERSUS EDUCATION

What most people think of as the entertainment industry—that is, the cinema and television aspect of the business—is based on a firm foundation of offering the public emotional stimuli so much more vibrant than the realities of everyday life that vast numbers of people will pay good money to experience vicariously assorted chills and thrills.

Audiovisual programs, in contrast, by and large are designed to provide material to inform, instruct, or change audience attitudes. So, inevitably, the question arises, "Why not add the emotion-stimulating techniques of the entertainment film to the information-providing, attitude-changing content of AV presentations?"

The answer, of course, is that it *is* being done. Emergency paramedical training programs often use simulations of a real emergency. Pilot training programs simulate conditions aboard a plane in flight. Interactive behavior modification programs present dramatizations of confrontations and let the viewer respond.

There are also a host of delightful animated videos that take viewers through the body's circulatory system or teach children how to respond to overly friendly strangers. Flashy, fast-moving video walls not only display and sell merchandise, but convey the message, "This company is on the ball—active, up-to-date, fun." MTV videos and rap shows entertain while they tell viewers—sometimes subtly, sometimes flat-out—"Cool dudes don't do drugs."

Do emotional touches and amusing bits waste time and undermine a proper learning atmosphere? Arguments can be made on both sides of the question. Particularly where technical material is concerned, clients and viewers alike may frown on anything that departs from a flat, strictly factual presentation.

I recall one instance where I devoted considerable sweat to incorporating humorous touches into a training program with a reputation for putting its viewers to sleep. When the firm's president saw the script, he promptly ordered me to remove all the comedy bits. He wanted facts and operational details and that was all. (Interestingly enough, he did have a sense of humor, and he did guffaw at some of my efforts. I never could figure out why he objected so strenuously to letting trainees have a chuckle also. All he'd say was that light touches would "distract" them—a statement with which I still disagree heartily.)

Be that as it may, the fact remains that there are those who want their information straight. Some professions are oriented to the brief, direct presentation of material with little or no embellishment. Snappy chatter irritates them. Doctors, scientists, and people dealing with statistics seem to fall into this category. That is not to say that they lack an ability to laugh, for they do have it. It's just that when they sit down to learn, most often they want the facts—straight, brief, and concise. They don't need motivational pitches or inspirational tidbits.

On the other hand, there are groups unwilling to pay any attention whatever to your program unless it's served up with a dollop of entertainment. In general, these are the involuntary audiences, the assemblages dragged in at the end of a rope. The sleepy students, the military recruits, the bored sales force that feels it's heard—and been put to sleep by—it all before. They have no prior interest in your subject, and in effect they challenge you to create one.

This often—though not always—can be done. It's possible, upon occasion, to build a story line into even a fifty-frame slide show or to add tongue-in-cheek touches to a museum display. Not many of us are interested in how to clean toilet bowls or high-potency deodorants or the latest antacid, but television commercials still strive to grab our attention while selling these products. Frequently, they succeed.

To such an end, it never hurts to try to incorporate physically interesting, imagination-stimulating objects or action into your visuals. A unique artifact in an archaeological presentation or a highly magnified bacterium shown on a slide with dramatic narration about the threat it offers may prove effective in the right place.

The big thing is, you *must* know your audience—not the easiest thing in the world to manage because of the large number of variables.

Variables—they can prove so treacherous, so confusing in any field. Thus, a brilliant young scientist I knew was having a terrible time over an experiment on which his whole future quite possibly depended. Always, when he performed the experiment in his own laboratory, he got exactly the same result. When others tried it in their laboratories, they didn't.

The scientist settled down to checking details. When at last he pinpointed the significant factor, he discovered it to be, of all things, the test tubes he was using. Trace elements in the glass were affecting the validity of his work. New test tubes and a change in method resolved the problem.

As a scriptwriter, you may upon occasion encounter somewhat the same kind of thing. Frequently, the issue is false assumptions. Too many people, for example, assume that an Indian is an Indian. They're wrong. When the furor about the American Indian Movement and the Wounded Knee incident was at its height, I found that many members—and not just older members—of the tribes involved were anything but happy about it. Not all women are feminists. Not all Catholics stand with their church on the abortion issue. This is why you, as a writer, *must* check on possible variables. Will your viewers include religious groups with dietary taboos you haven't considered? How about alcohol? Is national or regional pride or prejudice a factor? Has your audience experienced snow or sea or modern farm machinery? What's their reaction to credit or car ownership or extended families?

Nothing can replace a wide range of experience on your part in working out such matters. You need to make it your business to work with different groups, different cultures, and different age levels wherever possible. Beyond that, reading can help. Explore psychology, biographies, history, technology, sociology, and anthropology. These topics will broaden your background and your insight, and the things you learn will be well worth the hours spent.

## CASE STUDY 1

*How much difference does audience make? Consider the case of the twenty-minute film,* Guy with an Itch. *Produced for the National Science Teachers Association, its purpose was to raise the question of "What kind of people are scientists?" to seventh and eighth grade students. The answer, and residual impression, was pinpointed as "Scientists are people with an itch . . . curiosity-motivated men and women who are intrigued by the unknown and who tend to note and probe gaps in humanity's knowledge, and in so doing directly or indirectly bring progress."*

*Because of known negative (survey-determined) attitudes toward science on the part of the audience, determining angle and approach were vital. It was essential to replace students' stereotypes of scientists as super-intellectual, painfully sober old men with white beards who never leave their laboratories with a positive image of lively human beings attuned to reality and their fellows and, above all, curiosity.*

*The question was: How should this be done?*

---5---

# Ideas versus Media

Every program begins with an idea: Your idea. Your client's idea. A producer's idea.

It doesn't matter where said idea comes from. Only one thing is important.

The idea must excite someone.

Especially you.

Indeed, let's make that our working definition: An idea is something that excites you about your topic.

Which means that an idea may be the sudden flash of thought on a dull afternoon that says, "Hey, why don't I pick up Terry and go to the beach?" Or it may be the realization that you can end the chore of grass mowing forever by pouring the front yard in concrete and spray painting it green. Or it may be the impulse to duck the diet problem by buying a new skirt featuring vertical stripes. On a less frivolous level, your idea may be to suggest that your client build a resort promotion campaign around a colorful kachina doll collection.

Perhaps this will come into sharper focus if we consider the novel *Psycho*, written by my friend Robert Bloch, a master horror writer.

I scanned *Psycho* when the book came out. It read well enough, but I could see nothing in it to raise it above the level of Bob's other tales. Indeed, there were several others I'd have rated higher.

Then, along came Alfred Hitchcock. Next thing I knew, *Psycho* was off, running, and racking up multimillion dollar grosses.

What was the difference between Sir Alfred and me? I enjoyed *Psycho*, and that was about it.

Sir Alfred? Clearly, *Psycho* excited him as an idea. At some point it generated a sense of mounting, goal-oriented inner tension in him.

In the process, it started creative juices flowing—juices that, channeled through Hitchcock's genius, developed the idea into one of the top-grossing films of all time.

So much for *Psycho*.

On the other side of the fence, some years ago I wrote a science fiction novel titled *The Transposed Man*. It was ordered and paid for in advance by a magazine that had bought nearly half a million words from me and had never rejected a story.

The editor decided to make an exception with *The Transposed Man*. It came back to me like a yo-yo on a string, with a six-page letter that detailed its flaws and described revisions that would have made it a totally different story.

Well, I still liked—that is, was excited by—the original idea. So I declined the request for revision, refunded the magazine's money, and set about marketing the piece elsewhere.

To make a lengthy chronicle shorter, it sold. First, to an American magazine, then to a U.S. paperback house, then as a British paperback, then in Germany and assorted other countries. Twice, it almost was made into a film.

What lesson can we draw from this?

Simply that an idea is only as good as your own reaction to it—your involvement with it. Where and how it originates is of no account. Sometimes an idea hits so hard it jolts you out of sleep in the middle of the night. On other occasions it may pop up as an off-the-top-of-the-head response to an unanticipated phone call or unforeseen assignment. Or, it may be the product of routine brain racking as you search for a theme for a new industrial fair display or a means of setting forth dull but essential data more vividly. All that matters is that you scrabble through the farther reaches of your mind till you find that fresh and different something that somehow, indefinably, fits the situation at hand.

The other night a friend dropped by, virtually glowing in the dark. He'd been given a video to script on the dull, routine subject of safety procedures in case of a fire on an aircraft. Dreading the frustration of what appeared to be the inevitable rehash of manuals on the subject, he'd dreamed up a simulation of an A-7 in flight when a fire's discovered. The crew reacts by the book. Then the tower responds and, in its turn, the base's fire-fighting unit rolls out.

Later, at the hospital to which the crew's taken as part of standard operating procedure, the fire chief and the pilot discuss what happened and thus review the procedures. And the pilot is reassured to learn that the fire units each have a diagram of the aircraft that they check over while waiting for the aircraft to land.

Thus, the instruction on dealing with aircraft fires was handled in a livelier, more interesting way. My friend got a commendation for taking the initiative as he had, coming up with a fresh new approach—an idea— to bring life to a less-than-inspiring problem.

Enthusiasm is another vital element when you're working up ideas.

On one occasion, a producer phoned to ask if I knew anything about how highways were built.

Now among subjects that left me blank, road building would have come close to the head of the list. But automatically, I hedged a little, "No more than any average guy. But tell me what's up. I'm willing to learn."

"What we need," said my caller, "is a video about a road-paving machine. It's fascinating stuff—good visuals and all that. But it seems to turn off most of the boys."

"Not me." I was already beginning to get excited. "I mean, anyone who's ever read Ted Sturgeon's *Killdozer*—"

The producer clearly didn't give a hoot about my tastes in science fiction. What he did like was my obvious, instant enthusiasm. (I've been told more than once that my most valuable asset is my ability to get excited about any subject on ten minutes notice or less.) In this case I got the job and, to this day, whenever I drive past a road-repair crew I find myself checking out the equipment. So remember, always, that an idea is only as good as your own reaction to it and your involvement with it. Where and how it originates is of no importance.

How do you find ideas? Is there a universal wellspring from which you can draw excitement? Research is a good place to start. Dig into any topic in enough depth, be it dandruff or dumbbells, and odds are you'll turn up enough information off the beaten track to rouse your curiosity, interest, and ultimately excitement, and thus get you going.

What I call the *list system* is another gambit that works for many. When confronted by a project that calls for an idea, sit down and list possible approaches, good and bad, that relate to the project one way or another.

Note that I say, "good *and* bad." Why? Because at this point you must try to avoid judgmental thinking—that is, the tendency we all have to classify things automatically as "sharp" or "dumb," "right" or "wrong," "acceptable" or "unacceptable." You'll do this later, when it becomes time to sort the ideas out. But what you want first is *all* ideas on your topic, no matter how offbeat or crazy.

Indeed, this is what creativity's all about. Simply put, it's multiple responses to a single stimulus—forcing your mind to consider a variety of possibilities rather than just one. When I taught writing, students often complained that they couldn't get fresh ideas. Whereupon, I'd remove the doorknob and shaft from my office door and set them on the desk before the student.

"Make me a list," I'd say. "Tell me ten different ways to kill somebody with that knob. You're forbidden to leave till you've done it. If you can't, I'll understand—and you can drop out of the course."

Invariably, the first entry on the list would be "Put it in a sock and hit the victim over the head with it." "Sharpen the shaft and stab the victim with it" came in a close second. But by the time students got to No. 10, they'd moved on to needles injecting mysterious poisons, circuits wired to electrocute the hapless martyr, and explosives set to go off at a touch.

That's creativity. That's the key to finding ideas that are fresh and different. But that is not to say that fresh and different will always be all that fresh or all that different. Old ideas, refurbished, sometimes prove best of all.

Be that as it may. What counts is that, at whatever length and by whatever means, you track down an idea.

How do you know whether it's a *good* idea?

You try it out on other people involved in your project to see if it excites them too. If it doesn't, it could be it's back to the old drawing board.

Case in point: I had an absolutely smashing idea for a TV show episode. Off the top of his head, the story editor catalogued four previous occasions on which that idea had been the focal point of other shows.

Naturally, nothing like that ever will happen to you.

When you have an idea that turns you on, run it up the flagpole and see if anyone salutes. If someone does—a key someone with muscle— you're on your way!

Almost. There's still another step. You need to be critical and judge your idea's suitability for AV.

Since AV means your idea is supposed to be designed for sight and sound presentation, you need to ask yourself three questions:

1.   Can this idea be communicated effectively in AV form?
2.   Do other AV materials already cover this ground?
3.   Is anyone interested?

Premise: *Anything* can be communicated audiovisually. Boolean algebra. Relativity theory. Gnostic philosophy. Anything.

Effectively?

That's something else again. AV's best adapted to things you can see and hear.

Yet if we apply that notion literally, it limits us to topics like carpentry or musical notation or real estate sales. And we know that's not true.

The real issue, of course, is that every subject is potentially subject to effective audiovisual presentation if you can bring imagination to bear on it in such a manner as to *make* it seen and heard. So the issue really becomes whether you have worked up your idea in a manner calculated to fit it comfortably into audiovisual form.

(Don't worry just yet about *how* to do this. We'll take that up in later chapters.)

What about those "other AV materials" referred to in the second question above?

If you're considering a filmstrip or slide show on the life cycle of the duck-billed platypus and you discover that three other explorations of the subject already exist, it stands to reason that yours must offer some distinctive advantage if it's to sweep the field.

On the other hand, don't be too easily put off. Love stories in the thousands have seen print down through the years, but that doesn't prevent a host of writers from producing new ones. The issue is to add some new twist or fresh touch to your idea.

Consider the third question: Is anyone interested?

Anyone with the money or muscle to do something about your idea, that is.

A friend of mine decided that a filmstrip to be titled *Poisonous Snakes of North America* would prove a sure-fire winner. A number of herpetologists agreed it was a splendid subject.

A major distributor refused to touch it, however. His comment was, "I don't believe I'd sell six prints." Others echoed his opinion.

Crass commercialism, perhaps. But you do have to take it into account.

## CONTENT PLUS PRESENTATION

Assume now that you do have an idea that meets the tests set forth above. Immediately you're faced with a new dilemma: Precisely what do you want your presentation to say, and how best can you say it?

In other words, the issue becomes one of content and presentation.

This can be more of a problem than meets the eye. Why? The exercise is one of finding the proper approach for the audience. Your assumption that ulcers are psychogenetically induced, or that the reality of flying saucers is universally accepted, may not get the reaction you expect.

For that matter, even the basic information from which you're working may prove inaccurate.

A case in point: My wife once was employed to script a program intended to promote the work of a state's American Indian fashion designers.

Attacking the project with enthusiasm, she soon discovered a fatal flaw in the thinking behind it: Locating these designers was a major task. Few had telephones, no definite addresses were available for many, and most fell into the category of cottage industry. Only three could produce garments in quantity and deliver orders on time.

It was hardly the program the project's sponsors had envisioned. The piece ended up as a magazine article.

Similarly, where presentation is concerned, some of us like the light touch; others prefer profundity or its simulation. Straight chronological handling holds charms for many. But there are those who incline to a problem-solving or an inverted pyramid mode of attack.

These are matters we'll take up in much more detail in Chapters 6, "Concepts, Scripting, and You," and 7, "Developing Your Idea." For now, merely be forewarned that the issues do exist.

## SELECTING YOUR BEST BET

"Everybody wants a multiscreen slide show," says William E. Pryor, independent writer/producer. "Talk them out of it, if possible. Two projectors, a dissolver, and a sync playback unit are as complex as they should get. Because all that other stuff isn't portable. I made $8,000 converting a multiscreen slide show to 16mm film a couple of years ago because the guy who made the slide show never thought about how to transport eight projectors, etc., on a plane."

The point Pryor raises is valid.

A further lesson is that you should never overlook the conditions of use of the product. What works in a conference room with established groups brought in as viewers may bomb if it's carried to a variety of locations and shown in rooms of varying sizes, shapes, and acoustic qualities, with assorted electric outlets, light control devices, and such, and the janitor serving as projectionist.

But when so many factors are involved and when they intertwine and counterbalance to such a great degree, how do you pick the most effective approach? In the last analysis, it's a matter of becoming aware of variables and then weighing each against the others. In other words, painful as it may be, you have to make individual judgments. Yes, you may be wrong, and being wrong can prove disastrous. But that's the name of the game.

## CASE STUDY 2

*Finding an idea for* Guy with an Itch *was a bit different from many projects. That is, the content was predetermined and summed up in the statement, "Scientists are people with an itch," so presented as to reshape the student audience's stereotypes.*

*How should this message be presented, in general terms? What was the best way to reach the audience?*

*It was a time for nail gnawing and floor pacing. One thing soon became clear. The usual approaches simply wouldn't do. Let a real live scientist or simulation thereof step onto the stage, mouthing platitudes and telling the students what they "should" believe, and those already disgruntled and rebellious portions of the audience—that element the sponsors were particularly eager to reach—would hoot it off the screen.*

*Alternatives? The note pages tell the story. Acceptable ideas were hard, hard, hard to come by.*

*Only then, an idea—an exciting idea! Why not abandon reality entirely? Why not jump over into fantasy land—maybe even make use of caricature figures, kookier by far than any actual people, presented in animation?*

*For better or worse, and whether the product of normal thinking or sheer fatigue, it was accepted.* Guy *was on its way!*

*Film  No. 2*
*Title:*
  *Q:   What kind of people are scientists?*

*Residual impression: Scientists are
    people w/an itch ... people who
    are intrigued by the unknown.*
*Essential content: Establishment of
    scientists as curiosity-motivated
    types who tend to note & probe
    gaps in man's knowledge &, in so
    doing, directly or indirectly bring
    progress.*
*Approach: Via animation, & w/barbed,
    slangy narration, we view the rise
    of science in historical terms, from
    cave to grave.*

#2   Eyes in the Dark

Seg. 1 Guy w/an Itch

In seg. 1, our guy's itchg. in a cave. He
figures out that the prob. is fleas.
ignores his assoc.'s' laughter...
drowns the fleas.

A scientist is a guy who scratches
when he itches. Dog Smith.

Like Lemuel Q. Smith

Finally we reach the people whose itch is
all _inside_ their heads.
2   Example of a simple, primitive itch.
3   Etc.
7.   Finally we get to the modern scientist
& follow thru on him.

Play this for laffs all the way.
Be sure to introduce at least one woman.
Maybe do the whole thing in verse as
a hillbilly song?

Back scratcher - x - ?
Replace w/ jeevers?
Conclus. — too pat — kids shdn't voice
     approval or we'll create negativism.
Close on non-restg. place a la "duck
     may be somebody's mother." Don't
     give wind-up-recap. Take out
     honey.
Make use of antagonist — like someone
     jeerg. flea bit — "Look at that idiot!"

p.17. — synonym for "compulsion"

Do Archimedes verse over in exact
    chronological order.
Kookie? (stupid?)
Eliminate "ain't", "don't," other
    faulty grammar.

# Concepts, Scripting, and You

Like everyone, I have my pleasures. Among these is visiting art galleries large and small. In one such gallery, a few years ago, I encountered a unique presentation. As I entered, I found myself in a garishly painted anteroom or foyer. A continuing mural showed a range of jungle vegetation—trees, ferns, orchids, strange plants and grasses, and the like.

The dominant feature was a wavy line, clearly defined and perhaps four or five inches wide, which wove in and out through the landscape a bit below shoulder height (but on the ground, in terms of the painting's perspective). It extended all around the room, its ends reaching into an alcove that led to the gallery.

My companions and I were puzzled. After a few moments of trying—and failing—to make sense of it, we moved on into the alcove.

Here the line ended. On one side, it terminated in a fantastic serpent's head; on the other, in the serpent's tail.

We laughed, of course, the way you do when you've been put on, then proceeded—a bit warily—to view the rest of the exhibit.

Thinking back in the years since, I've come to wonder if what I saw didn't offer a bit more than the visual joke I had at first acknowledged. The artist had developed a concept: a meaningless line, given a head and a tail, became a snake and thus acquired meaning. It jolted the viewer into paying more attention to the display and looking for the unexpected. There were surprises to be found. The viewer became involved.

This same principle, properly applied, is the ideational foundation stone of any AV project.

## CONCEPT, CONCEPT . . .

I hope you'll forgive me if, despite their dictionary definitions, I view *concept* in a somewhat different light from *idea*. Although both spring from the mind, I see the concept as tending to evidence somewhat more bones and meat. To put it another way, the idea is an exciting trigger; the concept, to a greater or lesser degree, the developed product of the shot the idea fired.

How does a concept come into being? For me, at least, the process can be narrowed down to a fairly set procedure.

First, of course, comes the idea. *The idea, by its very nature, always centers on a topic.*

Thus, your assignment is to devise a fresh new audiovisual program for the natural history museum where you handle public relations. The topic: parasites in man and animal.

Attacking this assignment, you scowl and scratch and pace the floor, hunting for some fresh angle.

Lists are one of the writer/conceptualizer's most useful tools. When you need an idea, don't bother agonizing for the "right" one. On the contrary. Just make a list of all the notions—good, bad, indifferent—you can think of. Before you know it, said right one will turn up.

So. Parasites in man and animal. Rats, fleas, lice. A dreary index.

What other possibilities are there? Rats, fleas, lice. . . .

Tapeworms!

An idea has been born.

With the idea comes, quite possibly, the makings of a key point: a clear, concise statement of the precise thesis—the essential thought—you're trying to sell your audience. (Some call it a *core assertion*. Suit yourself as to terminology.)


## POINTING UP YOUR POINT

How do you identify the key point? Taking your topic and idea firmly in hand, ask yourself a simple question: "What about it?" The answer, when properly refined, is your key point.

"What about tapeworms?"

For no perceptible reason, your mind replies, "Tapeworms are faithful friends."

Weird! Can you prove it?

That remains to be seen. Meanwhile, however, you've acquired a springboard: a point of departure that captures instant attention. You've something with which to work—a concept to flesh out and develop.

Some key points are far out in left field: "Tapeworms are faithful friends." Most, however, are quite routine: "Here's a widget that will save you money." "Mothers' milk remains the best." "Play is children's work, a tool for growth."

Unless you search out and compress your project's message to some sort of summarizing statement, your presentation's unlikely to go in a proper straight line. "Define the statement of the film in one sentence," commands Stan Hayward, in his excellent list of rules for animation writers, in John Halas's *Visual Scripting*.

This advice applies equally well to every audiovisual form. If you can't nail down your key point in one sentence, you don't have your concept clearly in mind.

Difficulty in this area may come from a variety of sources. The most common one is that you're trying to work with a mixed bag—bringing in extraneous issues, meandering instead of marching. More about this in Chapter 7, "Developing Your Idea."

For now, let's assume you do have an idea and a clear, concise key point.

## WHY SCRIPT IT?

To become truly meaningful, a concept must be taken out of your head and communicated to others. Indeed, failing to do so, how can you hope to interest others in your concept? Faced with new thoughts or unfamiliar ideas, we all need some sort of detailing and description to appraise and ponder.

Such a description, in audiovisual, is termed a *script*. It sets forth your proposal in written form so that your associates may admire your brilliance. In the course of developing your script, your work becomes subject to discussion. Confusions are cleared up, obscurities eliminated, points of difference resolved.

You're going to produce the project yourself? Write a script anyhow. The process of putting your thoughts on paper will sharpen your presentation immeasurably. Reflection, study, analysis, and debate—whether with yourself or others—can't help but improve your work.

Beyond this, a script will bring other benefits to you and your colleagues—eminently practical benefits.

For one thing, a script simplifies production. The producer and director see what settings, actors, and action you have specified and so can plan accordingly. If scheduling is involved, it can be more easily arranged. If impossible or impractical features are called for, they can be pinpointed so that other, more feasible elements may be substituted. If contradictions or errors have crept in, they can be cut out.

In addition, a script cuts costs. Improvisation almost always proves expensive. Through preplanning at the script stage, production time—which is to say, money—is saved. Low-budget elements can be substituted for expensive ones. Duplications can be spotted; shortcuts, devised.

## WHAT KIND OF SCRIPT?

Scripts come in all shapes and sizes, as you'll see in Part Three, "Working in the Specialties." I've seen perfectly adequate scripts scribbled on the backs of envelopes—and others that ran to hundreds of pages of typescript.

AV constitutes a unique and quite specialized area. Writing for AV is very different from—and, at the same time, very similar to—other types of writing. It calls for a unique orientation that sometimes allows the writer considerably more leeway in choosing the pattern into which to cast the work than do most channels of communication. On other occasions, AV limits the writer more than other fields.

Sometimes, a simple present-tense narrative presentation is perfectly acceptable. In other cases—interactive or multimedia, especially—the instrumentality is so new and the techniques so complicated that no standard script pattern has yet been developed.

In most instances, however, scripts start from some sort of outline: a proposal or a treatment.

The next step may be a production script, in either a one- or two-column format.

Finally, there's a chance you'll want what's known as a *storyboard*: a graphic summary of the material in art or photographs. (The storyboard sometimes is prepared before the shooting script instead of after. It depends on circumstances.)

We'll discuss each of these ways of handling material in the chapters ahead. For now, suffice it to know that they exist.

All this may seem to be making mountains out of molehills as you contemplate the slide show you propose to put together about your vacation in the Poconos. But bear in mind that it's not only your prerogative but your duty to match script to project, and the big issue is detail.

As a scriptwriter, you'll set down your words for at least five publics:

1. Yourself—to clarify your thinking.
2. The sponsor—to nail down the message to be conveyed and the handling selected to convey it.
3. The producer and director—to spell out the presentation you plan.
4. The technical crew—to establish precisely what's to be seen and heard.
5. The ultimate audience—to create the effect you and the sponsor seek.

Back to the choice of script pattern. In planning your self-sponsored Poconos slide show—a program you'll also produce yourself—odds are that you seek only to set up an effective working order for the 35mm shots you made last summer. No producer, no director, no technical crew need be considered. As for the ultimate audience—a handful of friends and neighbors—you wish merely to share a pleasant interlude with them. You have no heartfelt message to convey, no product to sell. Therefore, back-of-envelope notations will perform the script functions nicely.

Suppose, however, that you have been hired to script an environmental presentation about strip mining. Slides and film or video footage will be shot in at least a dozen locales by people who will work independently. Extensive research must go into the project. Some sort of consensus has to be negotiated between the members of a large steering committee, several of whom hold almost diametrically opposed views. The results of all this must then be shaped into a finished project that will influence heterogeneous audiences nationwide to take decisive political action in the face of strong, well-financed opposition.

Does the problem of script become a bit more complicated than that faced in our Poconos idyll? And what can—and should—you do about it?

## GETTING STARTED

Let's talk about you. Long acquaintance with a wide variety of writers leads me to assume that you are human. That is to say, you come equipped with human attributes: strengths and weaknesses, virtues and deficiencies.

On the positive side, this means that in some degree you probably possess creativity, imagination, initiative, insight, passion, patience, perseverance, sensibility—oh, a host of elements.

The negative side? Here too you offer an amalgam of components, ranging from small pettinesses to a capacity for murder. But of all these, we need give heed to only one: self-doubt.

It goes by other names, of course: laziness, uncertainty, insecurity, vacillation, procrastination.

Perfectionism, even.

They all come down to one thing. You see it every day at swimming pools when someone hangs back, not quite nerved up to brave the water's chill.

The water, in audiovisual, is any new project. And the thing that makes the writer hang back is—full circle—self-doubt: the fear of laying an unsure ego on the line.

To put it in other terms, we all want to be right—every time. We want success, admiration, the plaudits that come with genius.

Sometimes we recognize that we can't be sure we'll win these. Especially when in new situations or when faced with unfamiliar problems, we can be wrong, we can fail, and we can look like absolute fools to our associates and to the world.

One way of avoiding such catastrophes is to dodge the issue—stall and put off and find excuses for not taking action.

That works, of course. The only problem is that it also condemns us to failure by default. Refusing to play the game, we never can win the championship.

There's only one cure, and I speak from experience: You have to seize the tiger by the tail and fight the job through—win, lose, or draw. Otherwise, experience, that dear teacher, can never tutor you for success. If, gnawing your nails, you hesitate, all's lost.

As an ex-cryptanalyst, I can tell you you'll sometimes obtain amazing results just by plunging into whatever morass confronts you and threshing around till something gives. In other words, you *must* act—even if what you do is wrong. The worst that can happen to you is that, missing your target, you'll check off one blind alley, one dead end.

## CASE STUDY 3

*An animated film about scientists for seventh and eighth graders. That's where the project stood. But that left it too broad, too general. What was needed was a key point, a core statement.*

*What was the film to be about? What was it trying to say?*

*Answer: Scientists are curiosity-motivated types who tend to note and probe gaps in man's knowledge and, in so doing, directly or indirectly bring progress. That was the message.*

*How to present it best—that is, most effectively—for a negatively-oriented audience?*

*In a humorous animated film!*

# 7

# Developing Your Idea

Through the years when I was operating within an academic framework, I had a succession of student secretaries—young women cashing in on their office skills to earn further education. Working with these students, I soon discovered that their most paralyzing fears centered on the writing of term papers. The envy they expressed at the apparent ease with which I did my own writing was flattering, to say the least. When I suggested that term papers really weren't all that difficult and offered to show them a simple procedure for dealing with them, their gratitude was heart warming to see—especially when they tried the procedure and found it really did work.

So, how do you write a term paper the easy way?

1.  Know your subject. That is to say, attend classes and study your text and carry out any additional research that's required.
2.  Pick your topic or, if one is arbitrarily assigned to you, work it up with proper care.
3.  Write down everything you know about the subject in a string of one-sentence paragraphs.
4.  Cut these one-sentence paragraphs apart and group them in terms of similarity, contrast, or contiguity.
5.  Arrange these sentence packets as logically as you can, going from problem to solution, the familiar to the unfamiliar, cause to effect, or any of the other patterns you'll find in a book on logic. Then, end with appropriate conclusions.
6.  Rewrite and polish as necessary in order to give the final product unity, progression, proportion, and continuity.

Congratulations on having acquired a new and useful skill for future reference!

These principles are equally useful in developing ideas for your work in the audiovisual media.

## BUILDING YOUR BACKGROUND

It's hard to write about nothing. Yet, perversely, that's what many beginners try to do.

May I suggest that, instead, you (1) recognize your need for information on your topic and (2) set about acquiring it in a reasonably systematic manner?

Our case study this time will be my wife's work in writing an interactive video program on the use of the oscilloscope. She started from scratch. Despite a better-than-average scientific background, she literally didn't know what an oscilloscope was.

Her first stop was *The American Heritage Dictionary of the English Language*, where she found a basic definition: "**oscilloscope** *n*. An electronic instrument that produces a visual display on the screen of a cathode-ray tube corresponding to some external signal."

The next stop was the technical manual for the particular model with which her script was to deal. Having read it, she understood at once why an AV program was needed. The manual, like so many of its kind, was totally beyond a layperson's understanding.

Fortunately, help was close at hand. A specialist had been instructing the sponsor's personnel in the oscilloscope's mysteries. Not only was he immediately available, he was eager to assist. Indeed, the main reason for putting instruction in AV form was that the workload had grown too heavy for him to handle.

Happily, the expert explained everything to my wife. It soon became clear that the difficulty was—as is often the case in such situations—that the expert was *too* expert. Years of experience led him to take all sorts of things for granted. He performed a host of steps automatically, jumping from short cut to short cut, unaware that in so doing he was bewildering his students.

Further, the oscilloscope involved was a highly sophisticated model, with a baffling variety of special controls. When my wife asked operators what function some of these performed, too often the answer came back, "Beats me. I steer clear of that one."

My wife, a persevering type, persevered. The day came when the program was finished and hailed by all and sundry—including the expert— as the clearest presentation of the subject they'd ever seen.

The point of all this is not my wife's triumph but the procedure by which she carried through the job. She began with the dictionary and the manual: *print research*.

The next step was probing the expert's knowledge: *interview research*.

Further study of printed matter followed, then work with the oscillo-scope itself—that is, *field research,* in which she applied the information she'd acquired to the actual manipulation of the instrument.

Only then did she get around to what nonwriters would call *writing:* the preparation of her script.

You might bear this in mind when you tackle your next job. A superior script is almost always built on a sound foundation of background study.

Although no set pattern for doing such investigation can be spelled out, certain broad guidelines do exist.

It generally helps to start with a quick look at what's been published on your subject. Don't do print research in too much detail, though, be-cause print tends to age fast and much of what you find may be outdated.

Now comes sifting through the experience of others—quite possibly experts: subject specialists, technical advisors. Properly handled, such con-versations may fill you in on all sorts of odd details unobtainable elsewhere.

Then, back to print research, if need be. And finally, out into the field, becoming familiar with radar, rodents, palmistry, pressure cooking, or whatever your topic may be.

Is this always the way you go about it? Well, hardly. Sometimes it's your good fortune to find yourself assigned an expert so eager and brim-ming with expertise that he or she gives you everything you need. In this case—congratulations!

(One word of warning, though: Be sure your specialist is as much a mastermind as he or she makes out. A couple of times I've been led down the primrose path by smooth talkers whose insights left something to be desired.)

Or, you may find everything you need in a clipping file, as I did once for a film on ulcers. Or, an examination of the widget or waffle or windmill may provide you with more than sufficient information. There are times, also, when you are presented with already-shot material that can't be duplicated. There's your material and you must work with that: for ex-ample, space probes, major celebrations, and nonrepeatable events such as earthquakes or assassinations.

However you handle it, though, be sure you don't get into too big of a rush. Indeed, where any kind of writing is concerned, it's wise to go slowly.

## MAKE HASTE SLOWLY

A major reason I prefer freelancing to steady employment is that it saves you from having to try to explain the strange ways of writers to employers.

One of the strangest of these ways, to the nonwriter, is our insistence that we're working when it's clear to all that we're just staring out the window. I well remember the difficulties I had convincing one of my former chiefs that he was getting his money's worth when I was sitting in a dark corner of a bar across the street from the plant with a glass of beer in my hand.

The point is that the human brain seldom works in neatly ordered patterns, especially where creative matters are concerned. Just as it takes time for trodden grapes to ferment, so a script's raw materials—ideas and research—must stew and seethe awhile before they effervesce and sort themselves into meaningful form. I've often been willing to swear I've explored every byway of a concept, only later to experience suddenly the thrill and exhilaration of having exactly the right touch flash without warning into my mind.

In consequence—and every experienced writer knows this—it doesn't pay to push too hard or to try to hurry things too much. More often than not, incubation time is essential when you're trying to shape jumbled thoughts into a script. Sleeping on problems helps. I still prefer to shave with brush, lather, and razor, simply because so many issues resolve themselves in those mindless minutes. A fellow writer walks his dog an hour each morning before he starts to put words down on paper. Allegedly, it's for exercise. Actually, he confesses, that period is when he gets his best ideas. Failing to provide for an incubation period, his thoughts become increasingly mechanical and routine.

On the other hand, don't fall into the trap of using creative brooding as an excuse for loafing—and believe me, that's easily done. You *must* keep it clearly in mind that you have a job to do, a script to write.

It helps, I think, to treat your brain as if it were a sort of cauldron. As you take ideas, copy, from it, it's essential that you pour in research—the raw material of creativity, certainly—and let it churn awhile. (For a brilliant study of this process, as applied to the poetry of Samuel Coleridge, see John Livingston Lowe's *The Road to Xanadu*.) Often, the infusion of research materials is enough to start this fermentation. Ideas applicable to your script come bubbling up instantly, with little or no effort on your part.

When that doesn't happen, give your mind-mash an opportunity to see the material unshackled. Forget the job. Take off and have that beer, see a film, or walk around the park.

Then, go back to your desk. Make a date with yourself to do so: "I'll knock off till noon (or three, or after supper, or whatever), and then I'll hit it."

When the time comes, keep your promise. Sit down with pad and pencil, word processor, typewriter, or recorder, and get busy. Odds are, the jigsaw puzzle pieces will fall into place.

## NAILING DOWN OBJECTIVES

Have you ever been lost in the Paris Metro system? No problem. The management has built in an interactive visual solution for you.

It takes the form of a big wall map strategically sited in major stations. Each line is indicated in a different color, and there's a button for each station. You push the button for the station to which you want to go. A row of lights flashes on, indicating the route you should take. All you have to do is get on one of that line's trains, transfer at the point the lights indicate, and before you know it you're at the proper stop.

What we have here is a crystal-clear example of the importance of objectives in AV work and how those objectives may be established.

Each project on which you'll work starts from an existing situation.

Now, enter the sponsor. He or she views the situation darkly and believes it needs to be changed in some way. So, the sponsor sets out to help make that change via an AV program.

Such proposed changes take many forms: "Children don't know enough about X. We want to give them information concerning it." "This piece of land is lying idle. We want to persuade people to buy pieces of it for homes." "These flowers are lovely. We want to show people how to grow them."

Or, in the case of the Paris Metro: "People are becoming lost in our subway system. We want a foolproof way to get them to their destinations."

Your problem, too, must have a sharply focused objective if it's to prove effective. So, how do you go about nailing down that objective?

First, you make very sure you know what your sponsor finds wrong with the existing situation.

Second, you ask the sponsor—and yourself—precisely what changes he or she wants to make.

Isn't this the same as the "key point" principle we discussed back in Chapter 6?

Not quite. A key point is a general statement of an attitude toward a subject: a point of view. An objective is a goal. It describes a new and different state of affairs you want to bring about.

Thus, "Tapeworms are faithful friends" is a perfectly accurate and acceptable declaration of a key point, presented with a wry, tongue-in-cheek touch.

Your objective needs to be more specific and goal-oriented in terms of some change you hope to accomplish via your AV presentation.

To that end—I recapitulate—you must check out the present state of affairs, ascertain what the sponsor finds wrong with it, and then spell out the change or changes that will make the sponsor happy.

In the example above, the sponsor holds to the view that a parasite is a life form that lives in, on, or at the expense of another creature.

The sponsor's objective is to convince viewers that parasites can be annoying, damaging, dangerous, and hard to get rid of, and that therefore viewers should do their best not to let such organisms move in with them.

The tapeworm then becomes a hook, a lead-in, a vehicle to help convey this message. Its importance is in no way diminished. But the key point that features the tapeworm is not the same as an objective, and it can be an unhappy mistake to confuse the two.

## CONTINUING ELABORATION

A major weakness of many beginning scriptwriters—and occasionally of professionals, too—is their tendency to delude themselves into thinking that a script springs into being full-blown. They couldn't be more wrong. A script is built a step at a time. And I use the expression "is built" rather than "grows" because ordinarily very real effort goes into the process.

That effort can be made a bit less painful by applying a process I call *continuing elaboration*. Continuing elaboration means just what it says: You get something down on paper and then elaborate on it—develop it more fully.

It helps if you go about this in a somewhat systematic manner. An approach that's worked well for me involves combining notes and type-script or print-out.

At the idea stage, I tend to scribble notes. These take in every aspect of my project that comes to mind, jotted down catch-as-catch-can with no effort at order.

Eventually, however, I bog down, or the jumble palls on me. Or maybe I just get tired and go away for a while.

Sooner or later, though, I return. And this time I take my place at my word processor and cast the whole mass of my notes into hard copy. Although I still make no attempt to organize the tangle, a certain amount of structure does creep in. Words and phrases become sentences. Sentences expand to paragraphs. Relationships of a sort appear. Pieces are juggled from one point to another.

The thing still makes no sense, you understand. It remains a mishmash. But it's a longer and more detailed and somewhat more coherent mishmash than it was before.

Again, in all probability, I take a break. It ends when I sit down with the hard copy and a pencil, crossing out, inserting, changing words, adding marginal notes, expanding.

The time comes when the printout is so marked up as to be next to unreadable. That calls for a return to the machine for clean copy.

But again, the copy changes in the process. Notions that seemed brilliant last night have now gone cold. New phrases pop up to replace old ones. Minor points take on new importance. Major issues lose their stress or fall by the wayside.

How long do you ride this seesaw? The only answer I can give is this: until you achieve a satisfactory—that is to say, satisfying to you—product.

This may not take as long as you imagine, for at some mysterious point your thinking synthesizes. Bored with groping, your subconscious suddenly shuffles bits and pieces into place. Pictures sharpen. A tightly-knit pattern of organization appears. The page that was hash an hour ago now marches neatly and precisely.

If you want a case study of how this can work out, take this book as an example. Its original plan was developed by this process, and so was each individual chapter and section.

So that's the continuing elaboration system. While it may not offer instant acclaim or pure perfection, it does bounce you off dead center with minimum waste motion and it does help you to move forward.

## HOW MUCH DETAIL?

Some writers write lean, sparse scripts.

Others tend to write fat, loose ones.

A third group favors what I call the *blank verse school:* prose poems in script form.

All have their proponents; all are in some degree successful.

My own feeling is that too fat a script, one developed in too much detail, tends to grow boring and so sacrifices attention. (I have been known to write too fulsomely myself, and I have had to watch sponsors, producers, and crews shuffle through my pages in irritation, trying to find the end to see if anything ever happened.)

Scripts that are too sparse, in turn, may on occasion prove ineffective, simply because in their brevity they don't get across a point or feeling. (Few of us read an outline or plot synopsis with the same enthusiasm with which we immerse ourselves in the original novel.)

"Poetic" scripts? Nowhere will you find a better field for such than in AV. "Mood" presentations may prove ideal for art, music, dance, and the like—anything where communication of emotion is the issue. A room bathed in deep blue light . . . a glowing orb or iridescent color against total black . . . sparkling Caribbean rhythms or Jean-Paul Jarre's futuristic music . . . ribbons of nylon net afloat in eddying air currents to caress the cheeks . . . —the sky's the limit! All you have to do is conceive it and get it down on paper. The right client will love it.

At the same time, remember that the *wrong* client will hate a poetic script. Excursions that are too imaginative seldom offer sufficient form to please sponsors who want tightly structured teaching programs or the equivalent.

Beyond all this, my own tendency is to try to capture the tone of a concept in copy—that is, the feeling and flavor you want the finished program to carry. When Cap Palmer's narration speaks of teen-aged Galsworthy Gulley "easing the pains of homework with the musical accompaniment of Malice Glooper and his All-Electric Aggravators," he's illustrating my point. So is the anonymous author of an industrial accident script who reports that "Harry catches his hand between the handle and the metal door sill with such force that he sustains severe abrasions to the skin and bruises the tendons and muscles. The injury is painful and Harry experiences difficulty in moving his fingers freely." Cap's goal is to maintain the light touch implicit in his approach to his material. The other writer strikes a more serious note—but again, one quite in keeping with the subject.

Finally, there's the matter of clarity. On shot instructions, especially, the crew needs to know what you're talking about, so for heaven's sake say what you mean, beyond all possibility of misunderstanding.

Whatever you do, do it in as few words as possible and as many as are necessary. Know in advance that experience will teach you a great deal more on the subject than anything I may write.

## CASE STUDY 4

*How was the idea of scientists as curious types best to be presented? Through historic incidents? Illustrative anecdotes? Biographical bits? All were considered.*

*The trouble was that such "creative adaptation" might of necessity deviate from fact far enough to arouse cries of outrage from the scientific community. It also could get heavy-handed and thus alienate key segments of the audience.*

*The solution? Abandon reality entirely, as in choosing animated film. Create farcical fragments that illustrated the point to be made painlessly, without rousing student antagonism.*

*The result? A selection of bits ranging from ridding a cave of fleas to discovering fire, from building pyramids to Newton's prism and a chart of the Copernican solar system.*

# 8

# Getting It Across

I knew a professor of archaeology who wanted a video presentation of his latest dig. It sounded like an ideal subject.

It ended in disaster.

Why? Because the professor, despite all protests, insisted on including lengthy discussions of the religious beliefs of a long-dead people.

A popular lecturer, skilled at playing to the audience, possibly could have gotten away with this. But the professor was *not* a popular lecturer, and the points he sought to make were both abstract and hypothetical. What he apparently hoped to do was to substitute an hour-long video for all of the anthropological and archaeological literature on his subject.

Believe me, this sort of approach won't work. Anything you can do to convince your clients that this is true will earn you a star in your crown.

As we've observed earlier, audiovisual combines sound and pictures. At its best, it offers an overview of a broad subject or a detailed analysis of an extremely limited one. Whatever you seek to accomplish with it, audiovisual operates most successfully when it adheres to a reasonable degree to certain rule-of-thumb standards. Just as a carpenter uses a hammer to pound and a saw to cut and does not attempt to miter corners with a brace and bit, so the scriptwriter helps the client to frame the message within the limits of the media.

## AN EYE FOR VISUALS

How do you present a smell on film or paper? How do you translate the word *creativity* into audiovisual terms? Can you devise a way to *show* me the atonality of Schönberg's music or to let me taste thick cream or chili peppers with my eyes?

And yes, such things can be done. Don't tell me they can't! After all, didn't someone record his interpretations of a sneeze in animation?

The point is, to become an AV scriptwriter, you must of necessity develop an eye for visuals: images that by themselves catch the audience's interest and tell stories. The more acute your eye, the higher your rating as a writer.

Further, developing such an eye may force you to expand your consciousness and stretch your imagination beyond your previous conceptions. Your thinking is likely to require readjustment to shatter the confines of traditional linear film or video.

Thus, to most of us, a visual is usually a picture or a graphic. But suppose your assignment is to create a walk-through exhibit encompassing five or ten rooms.

At this point, *visual* may switch in meaning from a graphic to light and color—all-engulfing, wholly or partially nonrepresentational, continually changing, and geared to set one or more moods. Perhaps there's audience participation also. Visitors push buttons, pull levers, step up or down or through light beams, peer into mirrors or through viewscopes, put on ear phones, feel textures.

Maybe the goal is to make the world of phobias, unreasonable dreads, real to those passing through. One room presents claustrophobia, fear of confined places; another agoraphobia, fear of open spaces; another, xenophobia, fear of strangers; another, nyctophobia, fear of darkness; and so on.

To achieve this end, a certain amount of construction may be necessary. Movable walls, for example, may be required; these too, in their way, constitute a visual. Indeed, such a program may even introduce you to a different kind of technical advisor, one who tells you whether the effect you seek is practical or within the budget's confines.

Do you begin to see why I say that a writer, in audiovisual, may be more a conceptualizer than a wordsmith?

Your tools on a job like this one will be unfettered imagination and a taste for variety. It also will do no harm if you're to some degree familiar with the artist's instruments—light, color, mass, line, form, proportion, contrast, rhythm, perspective, composition—as applied on levels ranging from realism to surrealism, actuality to abstraction.

Must you yourself be an artist? Not necessarily. But an awareness of art and graphics certainly helps. An excellent starting point is to study the work of others, both in AV and in other fields. Stay alert to all the things that visuals can do—the flood of ideas and information (and, sometimes, *mis*information) you can transmit through the eye. Salvador Dali has lessons for you, and so do M.S. Escher, Boris Vallejo, and Frank Kelly Freas. Maybe even Alexander Calder and Jackson Pollock.

Reading, too, has its place. You can do worse than to begin with Chapter 13, "Your Illusions Are Showing," of Magnus Pyke's *Butter Side Up!* Another top-flight stimulant is John Halas's *Visual Scripting*, particularly such articles as Samuel Magdoff's "The Advertising Message," Ernest and Giselle Ansorge's "Painter and the Moving Picture," Eino Ruutsalo's

"Typography in Motion," Roger Macdougall's "Writing Dialogue for Animation," and the Norman McLaren/Saul Bass piece, "The Experimental Film Maker and Designer." Roger Madsen's *Animated Film* and Raymond Fielding's *Special Effects Cinematography* are worth consulting for technical details on assorted procedures.

Beyond all this, however, the important thing to bear in mind is that audiovisual offers perhaps a greater opportunity for free expression than any other means of communication open to the writer. Many of your most frustrating limits, you'll discover, are self-imposed. The key lies in your own thinking. You can, on the one hand, restrict yourself to the flatly representational and pedestrian, walling yourself in with your own rules in concepts and visuals. Or you can search for the new, the fresh, the unfettered, and find yourself bounding through an exciting fairyland, a fantasia that delights you fully as much as it does your fascinated clients.

This holds true whether you're scripting slide shows or in-house video, interactive video or computer-based training, multimedia or dioramas. There'll always be those clients who insist on the dreary, the plodding, the program devoid of color or imagination. To make a living, you'll likely have to do things their way sometimes.

But there'll be others also, sponsors who seek something new and different. For them, the touchstone will be verve, originality, and unconventional vision. For their sake, and for yours, you daren't allow yourself to fall into moods of dullness, apathy, or inertia. Imagination is what paves the road to a reputation and a future.

Back down to earth. Audiovisual presentation takes a hundred different forms. At bedrock, every single one is the same. Once you learn the basic principles on which they rest, you can use all of them with adequate facility and master virtually any approach with minimal pain.

That's why I devote so little attention here to specific media. It simply isn't necessary. What you're after is visual interest, no matter what the format into which it's cast. That visuals should also convey appropriate ideas and information in an intriguing fashion goes without saying.

We'll take up details of describing visuals and the like in Chapter 10, "Writing It Down."

## AN EAR FOR LANGUAGE

More writers would script first-class AV programs if they'd learn to listen with both ears to good storytellers.

I base this statement on an experience from the 1930s, when the National Maritime Union was hard at it organizing Great Lakes shipping. To that end, they sent a host of deep-sea sailors drifting into Great Lakes ports, allegedly looking for work; actually, they were carrying the union banner.

Among these wanderers was a squat, foolish-looking Danish able-bodied seaman whose English was so thickly accented as to make you groan to see him get up to speak at the union hall.

You groaned the first time, that is. From then on you couldn't wait to hear him.

The reason was, Dane was a natural-born storyteller—a funny one, at that. From the first stumblingly hilarious anecdote, he was Seafarer Incarnate, and every man in the hall was with him. The facts he had to offer were nothing special. But when it came to life, spirit, drama—ah, there he had them!

Would it be treasonous to suggest that the narration in a good many audiovisual presentations would benefit from an infusion of life, spirit, and drama? These are all qualities valuable to narration. To put it in slightly different terms, straight telling too often doesn't tell much.

The storyteller's language was in keeping with his audience. Long words, by and large, were out. So were difficult words, complex words, obscure words. Active verbs, pictorial nouns, and shared experience were his stock in speech.

This should come as no surprise to any writer. If you're writing for farmers, you use farm terms and farm situations. For scientists, students, soldiers, or what have you, you do the same.

Too often, however, we let the element of life slip out. Spirit, drama, color—somehow an invisible pedagogue takes over and they fade away.

Thus, must the narration for a teaching filmstrip always begin, "This one-celled animal is an amoeba" or the equivalent? Mightn't the class pay more attention to "This creature is his own ancestor—and that makes him the oldest animal in the world!"

"You're not even breakfast size for a tyrannosaurus!" could have potential in a natural history display.

"Who do you hate the most?" lifts the voodoo doll out of anthropology and into life.

The point is, your goal in audiovisual is as much to influence and inspire as to inform. Changing attitudes is the issue, remember, as is influencing behavior.

That being the case, are you sure that narration and words are always necessary?

## A STING OF SOUND, A STRAIN OF MUSIC

Take the tyrannosaurus, above, for instance. Might it not be more effectively presented without words? How about a button for museum visitors to press that causes the tyrannosaurus to give a "realistic" roar. Attention thus captured, anything you have to say about it can be set forth on a placard or narrated later.

How about recording the sounds one hears within an Abrams tank? Couple it with appropriate jouncing movement and footage filmed through a gun port, and an audiovisual illusion is complete. Or let your audience take in the audio experiences that might engulf them in a nuclear submarine or a space shuttle.

And since we're speaking of sound, do you recall the paralyzing fright that erupts from a sudden crash of thunder as lightning strikes only yards away? Have you ever experienced the air of implacable juggernaut menace that accompanies the rumbling of an earthquake? For that matter, where would Wes Craven's *Shocker* or the endless versions of *Friday the Thirteenth* have been without their sound effects?

The same possibilities exist for your own work. The rattling of a rattlesnake, the ticking of a bomb, the hiss of air escaping from a tire, the slam of a door, the pop of a champagne cork—they're available when you need them. The radio sound effects specialists of the thirties and forties have charted the way for you in their amazing feats of mood and atmosphere creation and tension manipulation.

Music enters here too. Take movies. *Casablanca* tightened the throats of millions with the "Marseillaise." "We Shall Overcome," "Solidarity Forever," and "Dixie" rally the faithful to a cause. So, in a different way, do "Onward Christian Soldiers" and "Boomer Sooner." On records, the Beatles shaped a generation with "Strawberry Fields Forever," "Lucy in the Sky with Diamonds," and "A Hard Days Night." The Fine Young Cannibals' performance of "She Drives Me Crazy" and Michael Jackson's "The Way You Make Me Feel" carried a message that every so often translated into seduction. Depths of sensation first explored in Bartok's "Concerto for Orchestra" moved into the far future as Jean Michel Jarre probed through the world of synthesizers.

It would hardly be amiss to say that, to achieve its goal, a successful AV program often must go beyond what's thought of as its proper sphere and create an environment, a separate experiential world. Words alone won't do that. It takes sound and music too.

## THE ART OF ANALOGY

Too often, under pressure, writers fall prey to pedestrian thinking. We take refuge in a dreary literalism when we should be ranging high, wide, and handsome through imagination's farthest reaches.

One gate into these realms is that of analogy and metaphor: the drawing of parallels, actual or implied comparisons. Through them, we can bring the abstract, complex, and general down to earth and translate them into concrete terms. Walt Disney's *Fantasia* offers a delightful case in point, with its transmutation of music into the frolics of animals and imaginary creatures.

How better can you teach "Great music can be fun!" than this? Yet not one word on the subject has been spoken.

This kind of thing isn't limited to Disney. It's entirely practical for you to make the tool your own. Nor are elaborate measures necessary. The skull and bones—a symbol, a visual metaphor of sorts—spells danger to all. An insurance company's use of the rock of Gibralter as its emblem draws an analogy of stability. The gasoline ads that told you to "Put a tiger in your tank!" were drawing a parallel of power.

How do you get these parallels, these comparisons?

1. Decide on the point you want to make.
2. Make a list of things that symbolize or hold some analagous relationship to that point.
3. Pick the item or items that suit you best.

Put that way, the process seems idiotically simple. Yet when the cards are down, it really works.

Let's say you want to show soldiers the value of camouflage. Might it help to begin with such related examples as deer, snakes, snow bunnies, and chameleons blending into the background and then to show how a properly trained and accoutered infantryman can do the same?

Similarly, seeking to glamorize a resort, you might introduce slides of Acapulco, Cannes, and the Bahamas, drawing parallels to your spa and emphasizing how much cheaper and more convenient yours will be for nonjet-setters.

Does that seem too mundane? Let's go to heaven, then—a limited, stage-set heaven, played with a light touch and cast into an analogy of the resort.

Personification, particularly, is useful, since through it you can so easily create a prototype to fit your needs. In brief, visual and verbal metaphor offers a device that solves problems and helps to hold audience interest. Although you certainly shouldn't strain for it, you don't dare ignore it or its potential either.

## THE PROS AND CONS OF LOGIC

Rudolf Flesch once reduced all logic to two fundamentals: "Specify" and "So what?" (in *The Art of Clear Thinking*).

A generalized commentary on the tremendous power of storms does not carry nearly the force of a videotape of trees snapped like matchsticks and buildings shattered by Hurricane Hugo. And the "So what?"—that is, "What difference does it make?" or "Of what consequence is it?"—leads inevitably to the conclusion, "Humanity still stands helpless before the power of nature."

This is an approach that, to me, makes sense. But as audiovisual scriptwriters, we all need to move it a step further, via recognition that the things you *show* carry greater impact than well-nigh anything else. No number of words, no intensity of language, is likely to outweigh the right visual.

What can you say, for example, that will hit harder than footage of a napalmed baby or a dying AIDS patient?

Similarly, a shot of the great stone warriors guarding pre-Columbian Tula is more emphatic evidence of Toltec power than any words. Starving children in Ethiopia, their bellies bulging, bear horrid witness to the tragedy of famine. The hangman's noose and "Death for Drug Dealing" on posters in Malaysian airports punch home a message more clearly than any "Just say no" slogan. To countless thousands, the fall of Paris was represented by a balding Frenchman in a business suit with tears streaming down his face.

There may be a lesson here: Your best approach sometimes can be to establish mood and let your audience draw its own conclusion. Upon occasion, what's not said is stronger by far than what is.

Even when the issue isn't shock, but puzzlement, a picture that magnifies its subject—a fly's eye, a sugar lump—beyond recognition will catch attention. So will unique juxtapositions, unusual compositions of light and shadow, and many other things.

This is in no way intended to denigrate logic; assorted aspects of ordered reasoning are in considerable measure the topic of our next chapter. But symbolic visuals do constitute an often neglected element that's worth bearing in mind.

## CASE STUDY 5

*Every script focuses on decisions. Lots of them. In* Guy with an Itch, *one of the key decisions centered on how to tie the various elements together. Diffuse as the visual material was to be, narration came into focus as a central factor.*

*For example, was it to be objective or subjective, first person or third, sober or light and amusing?*

*Since the fantasy approach was already established, sober was almost automatically crossed out. But if not sober, what? And how could the various disparate elements be tied together?*

*Enter someone's stroke of genius: Why not introduce a semicomedic figure and center the narration in him? Make him timeless, as it were, and let him run through the entire film, linking the incidents that made the key point: Scientists are people curious about the world in which they live. Not only did it create a properly light-hearted mood, it could be developed in a manner that would offer the student audience someone with whom to identify and for whom to cheer.*

*Onward and upward!*

---------------------------------- *9* ----------------------------------

# Hanging It Together

A sales video, probably the most visually beautiful I've ever seen. The imagery, the flow of line, the flash of color—they caught you and held you tight. Truly, the cameraperson was an artist.

The trouble was that when the lights went on, you couldn't remember who the sponsors were or what they manufactured.

What had happened was that the photographer's artistry had seduced him or her. The photographer couldn't manage to subordinate that artistry to the assignment. An eye-catching composition took precedence over the sales message. Spectacular special effects rated higher than product identification.

Unfortunately, this artist's failing isn't limited to camerapeople. There's also the client who tries to crowd in unrelated detail, the writer who can't resist a clever anecdote or turn of phrase, the graphic artist who goes in for psychedelic bursts or "busy" backgrounds, and the sound recordist who lets mood music drown narration. One and all, they suffer from the same ailment: an inability to nail down the job they're paid to do and then build the project itself around it.

To paraphrase Cap Palmer of Parthenon Pictures, one of the finest scriptwriters I've ever known, the goal should never be an audience reaction of "What a brilliant film!" The note you should always seek is "Boy, are those great widgits!"

Or, "What a wonderful museum! I really must go there next weekend!"

Or, "So that's how resistor color coding works! For the first time, I really understand it."

Well, you get the idea.

## HOW ELEMENTS RELATE

One of the things that occasionally baffles would-be AV scriptwriters is that the AV field is so diverse. Consequently, no single approach fits all situations. No one solution solves all problems.

On the other hand, the categories into which presentations fall aren't all that different. The bonds that join them are every bit as marked as the walls that separate them. The thing that matters is knowing what you're about—the issues with which you're dealing.

First, presentations tend to take one of three major approaches; that is, they emphasize *instruction, mood,* or *experience.*

Not that these are mutually exclusive. Almost always, a degree of overlap is to be found between them.

Instruction is the most common of the three. Encompassing a high proportion of such forms as slide shows, filmstrips, teaching films, interactive videos, computer-based training, and the like, instruction tends to an orderly, logical approach.

Mood, in turn, centers on creating a feeling: happy, romantic, mysterious, excited, melancholic, or what have you. (Video walls are particularly good for establishing a mood.)

Finally, there's the experiential, where emphasis is on letting viewers "live through" a particular sensory involvement.

All three of these, please note, still center on what we originally said about audiovisual presentation in Chapter 4; that is, they change attitudes. It's merely the means they choose to do it that differs.

Thus, a video titled *Great Moments in Music* might offer pictures of famous orchestras, conductors, and composers, and its narrational comment might be interwoven with a background of taped symphonic fragments. Avowedly instructional, its approach could very well be direct and didactic.

Disney's *Fantasia,* in contrast, presents much the same content, but in terms of mood. On the surface, it's entertainment, pure and simple, but it too instructs. Implicitly it says, "Great music can be fun" and, as a correlate, "Great music doesn't have to be studied seriously."

The experiential approach might also be used with this topic. Perhaps the viewer enters a room equipped with one large and several small screens, a bank of push buttons, and volume controls. Each button is labeled with the name of an instrument. When the viewer presses a button, the large screen reveals a symphony orchestra in performance. Simultaneously, the instrument named on the button appears on one of the flanking screens; a performer playing this instrument, on another; the orchestra section of which it's a part (woodwinds, brass, strings, percussion, etc.), on a third. After a minute or so of solo performance, followed by a minute featuring the section, the orchestra as a whole comes on for a set period.

Again, this presentation offers instruction, but instruction of a different type and on a different level than the previous two examples.

How do you build such presentations, regardless of the category into which they fall?

You group their elements in terms of *similarity, contrast,* or *contiguity.*

As we pointed out when we discussed research in Chapter 7, the first step in organizing anything lies in putting like with like: birds with birds, mammals with mammals, reptiles with reptiles, or such.

Similarly, things *not* alike go in a separate class, that of contrast: big buildings versus small ones, meat versus vegetables, wood versus stone or metal.

A third division assembles contiguous bits; that is, those things that are next to each other, in the way houses stand next to one another on a street.

Thus, if your presentation is to be straight instruction, you gather the facts you want to put across, the vital information, pretty much discarding everything else.

(Why do I say "pretty much"? Because, as mentioned earlier, categories aren't mutually exclusive. Quite possibly it will seem desirable to salt in a *soupçon* of mood material, or some sort of experiential "audience participation" bit.)

To create mood, you may virtually eliminate facts, except as they contribute to feeling. At the same time, try to pinpoint precisely the mood that suits your purpose, the idea you seek to sell, then figure out how best to evoke it. Other moods, no matter how appealing, go by the boards.

(It also should be noted that distracting—that is, contradictory—moods inadvertently may creep into any kind of presentation unless you're careful. An anti-VD video fell flat because the woman on the screen was so attractive the men in the theater howled for her telephone number. The facial mannerisms with which Bela Lugosi terrified *Dracula* audiences when the picture was released now convulse viewers with laughter.)

An experiential program often will center on whatever cleverness you can devise to involve your public. The trick, as always, is first to make up your mind as to the program's purpose in terms of changing attitudes. Is the issue simulation of pilot problems? Demonstration of computer superiority to the human brain in certain areas? Testing mechanically for color blindness? Again, go back to grouping your elements as to similarity, contrast, contiguity. Add imagination, and work through to a presentation that does the job.

## ORDER AND ITS PRINCIPLES

The project is an in-house television program on minor industrial accidents. The key point will be "When your mind drifts, accidents happen" or maybe "Injuries come in when your mind goes out." The objective is to make employees aware that most accidents aren't accidental. (The ultimate objective is to improve efficiency and reduce time lost from work due to the many minor industrial accidents that employees too often regard as inevitable and the result of blind fate.)

What's our mood or tone to be? Because we'll feature relatively minor mishaps and because line workers tend to resent or be cynical about pontifical approaches, let's keep the handling light, and tongue in cheek.

The accidents themselves? We could use a shotgun approach and hit everything from shattered skulls to sliced shoulders. But that might make staging complicated and bring in too many serious injuries. We would do better to center on something more common and less critical, maybe a particular part of the anatomy—fingers, say.

This gives us, out of the blue and with apologies to the old *Laugh-In* TV comedy show, a tentative title: *The Fickle Finger*.

How do we open? Let's start with a montage of hands and fingers: fingers typing, sorting, pushing, pulling, poking, pounding.

Then, following the line laid down by our key point ("Injuries come in when your mind goes out"), let's show a series of incidents, blackouts, in which these fingers get into trouble. There's the stenographer who sees her boyfriend talking to another woman, so she hits a wrong computer key and loses a document. A box falls on the hand of a worker frowning over a racing form. A woman, upset because her mother-in-law is coming to visit, receives a bad cut.

The voice-over narrator, meanwhile, points out (with an appropriately wry touch) that fingers are faithful friends (there's that tapeworm line squirming in again), but they can betray us. That's because they're fickle and inconstant. They don't pay attention to business unless our minds do.

Finally, for a climax and to drive home the underlying seriousness of our message, a punch-press operator turns to call to a friend, at which point the press slams down, we hit a closeup of the man's face contorted in a scream, then dissolve to a matching closeup of his unhappy face at a later date, followed by a shot of a hand sans finger levering the press.

A recap of the earlier, less serious accident shots and maybe a repetition of the opening sequence of busy fingers brings our presentation to a close.

So. The job's done, the script written.

How do you organize such a project?

1.  Nail down a key point and objective for your topic.
2.  Choose a mood or tone to fit the situation.
3.  Pick an approach, a line of logic.
4.  Prove each point visually.

We've already dealt with the first two points. The last two warrant a bit more attention.

Thus, for point 3, there are a variety of ways in which you may attack any subject: chronologically, problem to solution, simple to complex, specific to general, general to specific, spatial, and a variety of others.

The approach you pick is one that provides proof that your key point is true, in such a manner as to move your viewers in the direction of the change or changes the sponsor seeks in the present state of affairs.

In our minor-accident example, we work from *specific instance* (individual mishap) to *general rule* (injuries come in when your mind goes out) and, within instances, from *cause* (distraction) to *effect* (accident).

In each case, also, in line with point 4, we prove each point visually. That is to say, we devise incidents that can be photographed. Why? Because, as some sage once observed, "The ear hears, but the eye remembers." What you see, you believe. When you back that with narrational interpretation and reinforcement, the impact is almost beyond resisting.

Is this the best way to present this subject? Not necessarily, for each script is unique and individual, and circumstances alter cases.

We might, for instance, very well have taken a *problem-to-solution* line, beginning with a series of accidents and then analyzing why they occurred. Or we could have approached the issue *spatially*, zeroing in on those departments that had the greatest number of accidents.

Whatever our approach, it's vital that we always bear in mind that we're trying to prove a key point and make a case for a particular core assertion. This demands that some sort of logic and visual evidence support our line of development. Always put your emphasis on *show* far more than *tell*. Narrational statements alone never are enough.

What about order of presentation?

Here you can learn from commercial TV; virtually every program starts with a "teaser": some intriguing action designed to catch viewer attention. Next, it builds through hills and valleys of pressure to a climax, a moment of peak tension in which mounting excitement rivets every eye to the screen. Finally, it ends with some sort of denouement that cuts loose the strain and ties up loose ends.

Our accident program follows this pattern. We started with action: busy, fast-moving fingers in a variety of situations. Each injury incident built from a low-tension "valley" of new characters introduced to a small "hill" of excitement as preoccupation brought misfortune. Together, they shaped into a profile of rising action that climaxed with the frightening, powerful punch-press episode. And, finally, tension was released and the lesson hammered home via recap and repetition of the opening.

Is this approach applicable only to in-house TV?

No, of course not. It works equally well with film (see my *Film Scriptwriting*, Chapter 3, "The Film Treatment."), filmstrip, slide show, video, or what have you. Space and time are your only limitations. Indeed, you'll note that our example could have been developed with equal effectiveness for a five-minute presentation or a fifteen-minute one. Even interactive segments should, when fit together, build.

## HOOKS AND HINGES

*The Fickle Finger*, you've no doubt noted, was comprised almost entirely of illustrative anecdotes—incidents that make a point.

Their importance to the scriptwriter and the value of being able to produce them on demand can hardly be overestimated. They provide hooks to lure viewers into subjects they ordinarily think of as dull, they act as hinges in changing from one segment of a presentation to another, and they form the backbone of many a program.

How do you discover or develop these handy items? A good part of the trick lies in the habit of keeping an eye out for them, both in research and in your thinking. Any time you read a biography or probe the evaluation of a process or product or idea, you need to be watching for those bits of action that can be cast into graphic form to help you lead on or manipulate your audience's thinking.

Where creating hooks and hinges from scratch is concerned, your task grows a trifle more difficult—history no longer provides you with ready-made twists and punchlines—but it still can be managed.

The list system, described in Chapter 5, constitutes your best tool. Ten minutes with a scratch pad gave me the digital details for *The Fickle Finger*. The list system needs to be coupled with a process I call "thinking through," however. You might also term it *projection*, or *extrapolation*, or even, perhaps, *reducing to absurdity*. All it really involves is asking yourself, "What *might* a given person do under these circumstances that would produce the effect I need?" Extend this process as far as your imagination will carry it via the list system, and you'll find it almost always gets results, as witness my own experience some years ago when I needed to create a sloppy character who would instantly be recognized as such.

Now sloppiness is, in its way, a difficult concept to present on film. The mere fact of dishevelment or slovenliness or soiling may indicate nothing more than being caught in the rain or having to spend a night on a park bench or in a ditch. True sloppiness is harder to capture in pictures, and I'm still proud of my solution.

What I came up with—after some hours of sweating and cerebration—was this: My woman is sitting at a table. A second character, approaching with a plate of scrambled eggs, trips. The eggs slide off the plate into the woman's lap. Momentarily—but only momentarily—she's taken aback. Then, not at all discomfited, she picks up her fork and starts eating the mess out of her skirt.

Note the mental process here. I considered a variety of possible ways of portraying sloppiness, rejected most, and finally picked one that made the point, even though it was a bit ridiculous.

## TESTING YOUR PRODUCT

Certain qualities characterize every effective script. While their presence won't necessarily guarantee success, the absence of one or all of them definitely does indicate a point or points of weakness. What are these qualities?

1. Unity.
2. Progression.
3. Proportion.
4. Continuity.

*Unity* means simply that a good script should be all of a piece, addressing itself to a single central idea or theme.

Disunity most often results from too many cooks stirring the soup. For example, the sponsor needs a video on pest control. The writer scripts it brilliantly. But then the engineer steps in. As long as the show's to be produced, wouldn't it be good sense to include a few shots of the company's product? The sales manager agrees. After all, every employee is, in a way, a potential salesperson. Why not have a little footage emphasizing how the product fits into every decor and color scheme? "Yes, yes!" cries the purchasing manager. Here at last is a chance to show the company's far-flung network of suppliers.

Well, you get the idea. What started out to be a first-class little show, unpretentious but effective, now ends up hash.

I don't mean to put all the blame on others. We writers are ourselves quite capable of being seduced by the extraneous.

Please, please, nail down your key point, your objective, and your audience before you start, then stick with them. Comedy relief is something you plan, something in keeping with the rest of your presentation—not cartoon gags stuck in just because they happen to be handy.

*Progression* might be termed the *element of forward movement:* the fact that you don't just reiterate the same point over and over. A good script is like a wall, built a brick at a time until the whole is completed. Your goal is a particular picture. But it takes shape only as you fit in the various pieces.

There's an important point to remember here, especially in teaching scripts: Any time you leave out a bit of essential information, it's like failing to put in all the treads and risers when you build a stairway. The student will bog down, just as the climber would fall going up the stairs.

To see the element of progression well handled, visit Amsterdam's Anne Frank house. Moving through it room by room, viewing relics and still photos, you end up with a deep feeling for the Frank family's plight.

*Proportion* is, I think, reasonably obvious. It means that your script emphasizes the things you want it to emphasize. The places scripts most often fall down in this regard is beginning, middle, and end; and no, I'm not joking.

Suppose the writer builds the beginning so big and powerful that it outweighs everything else.

Or, in the body of the presentation, the writer realizes that there isn't enough time or space or frames or whatever to develop it properly, so the writer leaves out vital data as he or she hops, skips, and jumps from high point to high point.

Or the writer comes to the end, and doesn't know quite how to wind it up. He or she tacks on three or four conclusions, one after another, and viewers find themselves irked and frustrated with a sense of anticlimax.

Solution? Plan thoroughly in advance, and allow time for rewrites as necessary. Any script can be given proper form if you work it through.

*Continuity* is merely a term to indicate that your script—and, hence, your presentation—hangs together.

To a considerable degree, the thing that *makes* it hang together quite possibly may be narration. Words spoken or set forth on title cards or wall posters can perform miracles when it comes to bridging gaps, changing subjects, and pointing viewers in new directions.

The thing to look out for is jerks or jumps—the sense of momentary bewilderment that assails the audience as it enters a new sequence. Step-by-step development, illustrative anecdotes, parallel constructions, narration, music bridges, and stings of sound to mark a change; all will help. But your main tool remains your own clear eye and straight thinking and your awareness that continuity is an element for which you need to strive.

Check your script out consciously, point by point, for unity, progression, proportion, and continuity. You'll still miss sometimes, but not as often as you would without that checking.

## CASE STUDY 6

*A script isn't just written; it grows. New ideas are incorporated as it goes along until the original concept is quite possibly changed almost beyond recognition.*

*In* Guy, *a twist virtually took over. Instead of normal straight narration, it was decided to cast the message into the form of a tongue-in-cheek pseudo-folk song, with guitar accompaniment and occasional sidebar interludes featuring boy-girl dialog in keeping with the picture's overall tone.*

---10---

# Writing It Down

A young man I know decided he wanted to be an AV writer. Personal contacts got him an assignment. The job took him into a shop where a thirty-year veteran also labored. Day after day, the veteran blithely came up with fresh ideas, new concepts, clever presentations, and copy that sparkled.

All the while, the young writer continued to sweat on his door-crashing first assignment. If he made progress at all, it was minimal. Finally, he couldn't stand it any longer. He addressed the old hand. "Why is it," he demanded, "that when you do these things, they look so easy? But when I tackle them, they're so hard?"

The veteran was a kindly man. "Paul," he said, "they're not easy for me either. But I've got thirty years on you, plus patience."

He was right, of course. Experience does count, and so does what an old cowboy phrase calls "staying with the cattle"—sticking with a task, no matter how hopeless it seems or how frustrating it becomes.

This quality of patience, persistence, perseverance—I don't know whether you succeed as an AV writer because you've got it, or because, as a writer, you develop it. But I do know you have to have it.

Beyond this, however, a system does help, especially when it comes to getting words onto paper.

A first step is to select your medium. What will do this particular job best for your client, considering everything from cost to conditions of presentation? Is a slide show the answer? A video? A multi-image presentation?

That done, your best approach is to have some sort of set procedure to follow.

This doesn't necessarily mean you do a lot of writing, however. Many projects involve more talk than typing, and your sponsor or producer quite possibly will be willing to forego any formal, step-by-step outlining of the project.

Nevertheless, for your own sake, to insure the clarity and precision of your thinking, you should work out your key point, concept, handling, and the like on paper, even if only in note form.

## IDEAS, INCORPORATED

My early days as a scriptwriter saw me develop an almost paranoid orientation where project proposals were concerned. I encountered much vagueness among other writers when I tried to find out how to write proposals properly. One told me one thing; one, another. "You just do it" tended to be the most common—and for me, most frustrating—response.

Time passed. Slowly, by trial and error, I discovered that my associates were right. A project proposal was indeed something you "just do." Why? Because each project is to a degree different from all others, with its scripting pretty much a law unto itself. No standard approach or handling or format exists. Each writer develops a technique he or she tends to fall into. Yet it's hardly the kind of scheme a writer can recommend or pass on to someone else.

Still, everyone needs someplace to start. Let's begin with definition. A project proposal, within this book's frame of reference, is a written recommendation that a specific AV presentation be produced. Designed to convince a potential sponsoring authority of the project's value and feasibility, it combines a summary of presentation content with such other data as the writer or producer feels may help to catch the sponsor's interest and persuade him or her that the project is important, desirable, and practical.

The heart of the proposal, to my way of thinking, is a statement of a concept and its associated key point—its core assertion—all elaborated into succinctly persuasive form.

(A concept, remember, is a topic plus an idea. A key point is a declaration of your attitude or point of view where this subject is concerned: a clear, concise avowal of the essential thought you propose to sell your audience.)

Proposals fall into two categories: solicited and unsolicited.

Solicited proposals are those submitted in response to an invitation to bid on a project. A government agency, for example, may decide it needs a particular AV presentation and sends out notices to writers or producers asking that those interested give ideas and prices. If you're a corporate employee—a writer for in-house TV, perhaps—the management may ask you to work up a possible approach for a desired show.

The unsolicited proposal? Since the writer or producer involved is attempting to create a market in most cases, it may offer a much stronger sales pitch than does the average solicited bid. That is to say, it tends to radiate what the writer hopes will constitute contagious enthusiasm for the project.

John J. Jones
1234 Medford Plaza
Crossroads, Texas

Prepared for:

CHAMBER OF COMMERCE
Crossroads, Texas


Project Proposal

FOLK ART, CROSSROADS STYLE

(Working Title)


The Crossroads Folk Art Museum's annual festival draws huge crowds and rouses tremendous community enthusiasm... breathes new life and vigor into our regional culture. Its only real weakness is that it lasts so short a time. A week, and it's over.

Couple this with the fact that so many tourists traveling through Crossroads, and so many residents in outlying areas, have no chance to enjoy the festival's color, excitement, and educational benefits. It adds up to a very real loss, both to community and visitors.

This needn't be. Simply, quickly, easily, economically, the festival can be turned into an exciting audiovisual program that will publicize our community, promote our museum, and enable citizens and tourists alike to enjoy our Folk Art Festival all year long.

Here's an unsolicited project proposal for a slide show. Note that at this point John J. Jones, our imaginary AV writer, is working almost entirely in terms of persuasion. His presentation is designed to sell an idea, a concept, a core assertion.

Specifically, what I have in mind is a hundred-frame color slide show, backed by tape-recorded narration and appropriate music.  Included will be shots of paintings and sculptures...artists and visitors...entertainment features... and the general setting/atmosphere characteristic of the affair.

The key point it all makes will be: "You'll have a great time--a _different_ great time--at the Crossroads Folk Art Festival."

The advantages of this approach are obvious.  Since we're talking about a slide show, it can be kept up to date at minimal cost simply by adding new shots of art, artists, and features... eliminating those which are outdated.  And narration can be changed for the price of a new tape.

Uses for such a program are numerous.  For example, it can be used to promote the annual show in other communities throughout our area.  Schools will find it invaluable in helping to build student participation in the Festival's junior division. A mailing about it to art groups outside the state will bring national attention.  Tourists visiting our museum during the 51 non-Festival weeks of the year will get a taste of the affair and so be stimulated to come back and see the Festival live. Such a list could go on and on.

From the Chamber of Commerce standpoint, this constitutes an ideal public service project.  All Crossroads will benefit-- and the Chamber will get the credit.

Tentative budget?  I estimate $3,500 should take care

Festival Proposal--3

of script, talent, production--the works.  And since the
museum already has its own cartridge projector and sound
programmer, purchase of this equipment will not be necessary.

     This is an opportunity Crossroads really shouldn't miss.
The profits the show will bring to the business sector alone in
a month will far outweigh the small investment.  And Crossroads
as a whole will receive cultural benefits beyond price.


Approved:

For Chamber of Commerce _____ Date _____

For Producer _____ Date _____

Solicited or unsolicited, the two things that count the most in a proposal are clarity and assurance. For this is a form in which understandability and self-confidence run neck and neck.

It also helps if, as an aspect of clarity, the proposal is typographically attractive. White space, subheads, indentation, and underlining make a proposal easier to read and comprehend and, quite possibly, approve.

Beyond this, there's little definite that can be said about proposals. Some are short, some long; most probably fall between two and ten pages. Some are written sparsely, some go into elaborate detail; it should be noted, though, that too fancy handling may tend more to confuse than to impress. Some proposals are flatly technical, some put emphasis on mood or emotionality or drama. And some lapse over into the treatment form, a topic that we'll now consider.

## TREATMENT: STEPPING SCRIPTWARD

A treatment is a concise summary of how your project's content will be developed. Written in the third person, present tense, and expository/narrative form, it spells out in greater detail the concept you set forth in your proposal.

It is, in other words, a highly specialized outline—a selling outline, for it emphasizes dramatization of material and communication of enthusiasm to the sponsor. Having read it, the sponsor should not only know more or less what the project's going to show but he or she also should feel at least a bit of the excitement and fascination with which you hope the ultimate audience will pulse.

How do you write a treatment? The best I can give you is a possible procedure:

1.  Record the informational elements you propose to incorporate in your project, setting them down in chronological order of presentation. By "informational elements," I mean factual content—the essential data to be included, whether it be historical events, biological processes, machine operations, or whatever.
2.  Record the interest elements to be developed—again, in chronological order. "Interest elements" means any matters of mood, feeling, or drama you want to become apparent.
3.  To the best of your ability, interweave these two aspects of your proposed program in such a way as to give the sponsor, reading, the same experience he or she would receive as a member of the ultimate audience.

Will this be easy? Hardly. Any old hand will tell you that developing an effective treatment is as difficult a job as any writer can attempt. But it's a skill you must learn. Without it, you'll never pass beyond apprentice level.

John J. Jones
1234 Medford Plaza
Crossroads, Texas

Prepared for:

CHAMBER OF COMMERCE
Crossroads, Texas

Project Treatment

FOLK ART, CROSSROADS STYLE

(Working Title)

Our opening slide features a closeup of Gus Faber's
famed "Crossroads Cowboy" folk sculpture.  Musical background:
"Keep Comin' Back to Crossroads" (Ken Casey's Crossroads Combo).
Successive slides present titles (OVERLAYS):

The Crossroads

FOLK ART MUSEUM

and

Crossroads

CHAMBER OF COMMERCE

present

These two pages of project treatment elaborate on the proposal and spell out in more detail the idea the proposal sets forth. By introducing specifics, telling what happens step by step, Jones paints a picture of the proposed show with words. Tone is set with "color" words, descriptive touches. In other words, Jones is still selling, as well as telling.

FOLK ART,

CROSSROADS STYLE

(Other titles as desired)

Changing angle, camera hits a MEDIUM SHOT that includes
both the "Crossroads Cowboy" sculpture (on workbench) and Gus
Faber, worn shirt and Levis, shaping up a new piece with hammer
and chisel.  Music goes down to background and narrator (GUS
FABER) comes in: "Yep, folks, that's me, all right.  Gus Faber,
workin' on a new piece to top my old 'Crossroads Cowboy', right
here at this year's Crossroads Folk Art Festival."

Cued to "Crossroads Folk Art Festival," we flip to a series
of slides that reveal the festival in progress...first in terms
of street scenes (especially Sutter Street, with its huge trees
and beautiful old homes) approaching the museum...the museum
exterior and the lawn show...the interior and Festival proper.
Gus comments on each new slide in his inimitable, wryly humorous,
down-to-earth fashion.

Featured slides will include shots of Ed Rogers and such of
his paintings as "Raspberry Roan," "Doan's Store," and "Bob-Wire
Gal"... Sam Hamilton's branding-iron candlesticks...Laura and
Effie Willis with their quilts...Sam Roper tooling leather...
Miguel Gutierrez blowing wine glasses...Roma Clarke and her
needlework...Wanda Elder with her wheeling hawk paintings...and
so on.

Visitors will be prominent in many of these slides, talking
to artists and artisans, inspecting paintings and craft

## SCRIPT AND ITS FORMATS

Let me reiterate here a point made earlier: In scripting an AV presentation, it is quite possible that you will never write a script at all in the normal sense of the term. The field is so fluid, the range so wide, the media so mixed that "standard" handling frequently turns out to be either inadequate or too restrictive.

Take the case of an experiential program. A simple narrative description of what happens to the audience, event by event and step by step, may be all that's required. An exhibit emphasizing mood may focus on sound or lights or a display's contents.

I doubt that the evening park show in the Michigan town where I grew up ever was scripted much beyond the level of "Fountains leap high into the air. Water cascades over the artificial falls. Varicolored lights, synchronized to background music, ebb and flow through the entire spectrum." But someone programmed the lights: red lights on three seconds, follow by blues and fountains, all on high three seconds, then lights numbers 3, 5, and 7 drop to low, etc.

Still, a script does help to keep things straight. This is the time and place to learn how to put one together.

You'll encounter three basic formats: the two-column, the one-column, and the master scene. (We'll ignore interactive and computer formats until we reach Part Three, "Working in the Specialties.")

In each case, any success you achieve will in large measure rest on two key factors: (1) your ability to visualize and see pictures in your head and (2) your skill at describing these pictures so clearly that production personnel can capture them on film or videotape or whatever.

Visualization involves a talent all its own. The trick is to learn to see, in your imagination, the individual pictures that ultimately make up your AV program. Sometimes, as in a slide show or filmstrip, these will be single photos. In other cases—film, linear or in-house video, some aspects of multimedia—motion pictures are involved. Again, the issue may be a walk-through exhibit, a display series, a mood or experiential presentation.

Whatever the medium, your job is to imagine what your audience will see or experience. In effect, you close your eyes, visualize your setting, view what happens there as your show progresses, and then start hitting your word processor's keys, describing the scene and experience.

How do you put those pictures onto paper? We'll take that up shortly.

It's not enough to look just for single images, however. You also need to watch for relationships between those images, so that your finished package will tie together neatly rather than jump about in all directions.

Such relationships can be established in two ways: A visual link may be set up between them, as in the cutting of a motion picture's continuity sequences, or it may be created through narration, as in compilation sequences. (These subjects are discussed fully in my *Film Scriptwriting*.)

Unless you propose to make a specialty of film, however, your best procedure is simply to spend as much time as possible watching effective presentations of the type you propose to script. Don't run each just once. A dozen times will serve you better. Note how, frequently, a shot seems to flow from the one preceding. This sometimes occurs because the second shot is of the same subject, but larger or smaller or from a different camera position, or because it follows continuing action.

On other occasions, markedly different pictures will be tied together by what the narrator says. Both means of relating material—visual and narrational—may also be used together.

The script samples appearing in Part Three of this book will help sharpen your awareness of these procedures.

But on to formats. Since the two-column is most commonly used, we'll start with it. On the left side of the page, it describes the material the viewer is to see. Anything heard—narration, dialogue, sound effects, music—appears on the right, roughly parallel with the visual content it's to accompany. The visual content is broken down into individual shots: that is, still pictures or a series of pictures taken in the same run by a motion picture camera.

The one-column format contains the same material as the two-column, but presents it with both visuals and sound (often termed *video* and *audio*) in a single column. This column may be set on the left, like the video side in the two-column script, it may be in the middle, or it may extend across the entire page.

The master scene script extends full page width. It differs from the other two formats in that it simply tells what the audience sees and hears, making no effort to break its content down into shots. Extremely popular in the entertainment film industry, it's very seldom used as a script form for AV presentations.

## WHAT ABOUT NARRATION?

An important aspect of any script—well, almost any script—is narration: words spoken (generally voice-over) to emphasize, complement, or supplement a program's visual portion.

When I so describe or define narration, I hope it comes through that, in AV, video (what's seen) by and large dominates audio (what's heard). As much as possible, the visual elements should tell the story. Narration is based on and cued to the graphic aspect.

At the same time, note how I've hedged: "almost any," "by and large," "as much as possible." Why? Because, upon occasion, you'll find you have no choice but to resort to what's termed the *illustrated lecture* approach: narration carrying the ball, with pictures thrown in to "emphasize, complement, or supplement" the audio—an exact reversal of what I said in the beginning.

```
                                    Festival script--11


    71   MLS, in food booth area.      Hey, now!  Art or no art,
                                       folks do get hungry, don't
                                       they?  Well, Crossroads is
    72   MCU, happy-looking visitor    ready for 'em.  Come one, come
         couple.  Man's pointing       all!  It's chow time!
         off slide.


    73   MLS, couple from 72.          MUSIC:  Mariachi group with
         They're headed towards        strongly Mexican number IN,
         food.                         UP, and DOWN to BG.


    74   MCU, tamale booth.            The tamale table's great, if
                                       you're a border type...

    75   MS/MCU, fat, happy Mexican    Even if you're not.
         in chef's outfit dipping
         into tamale pot.


    76   CU, plate of steaming
         tamales.  They look great.


    77   MCU/CU, 72 couple.
         Woman's taking bite of
         tamale.  Man holds a
         partially consumed tamale     MUSIC:  Combo (strong on
         ...looks delighted.           bones and banjo) with Dixie
                                       IN, UP, and DOWN to BG.
```

Two pages from the production script of *Folk Art, Crossroads Style*. Ideas are broken down into individual shots (slides) now, complete with narration and music. Note that Jones isn't content with flat, factual description. Visuals and narration alike come through in terms of tone, the effects our writer wants to create. Jones will probably have to rework many details after shooting's completed to make everything fit together smoothly. It's all part of an AV writer's job.

Festival script--12

| | | |
|---|---|---|
| 78 | MS, barbecue booth... different couple. Black woman (Delia Walker?) is serving. | Barbecue ribs, Crossroads style. Old South eatin . You-all don't know how good ribs can be till you taste these. |
| 79 | CU, different couple at booth. Angle favors man as he bites rib, sauce dribbling down chin for laughs. | MUSIC: Combo with <u>Deep in the Heart of Texas</u> IN, UP, and DOWN to BG. |
| 80 | Frame-filling CU, steaming bowl of chili. | Tex-Mex chili.   The flavor folks dream about clear 'round the world. |
| 81 | MCU, Adolph Fredericks in chef's cap at chili booth ...giving bowl to visitor. | |
| 82 | CU, visitor's enraptured face. | |
| 83 | Facial CU, Adolph.  He looks ferocious | That chili...Men have fought an' died to protect their private recipes. |
| 84 | MCU, Adolph.  He grips a six-gun stuck in belt. | |

```
                                  Festival--11


            INT. - EXHIB. HALL - DAY
     71     MLS, to take in food booth

            area.


            GUS: Hey, now!  Art or

            no art, folks do get

            hungry, don't they?  Well,

            Crossroads is ready for

            'em.  Come one, come all:

            It's chow time!


     72     MCU, happy-looking visitor

            couple.  Man's pointing

            off slide.


     73     MLS, couple from 72.

            They're headed towards

            nearest food booth.


            MUSIC: Mariachi group

            with strongly Mexican

            number IN, UP, and DOWN

            to BG.


            GUS: The tamale table's

            great, if you're a border

            type.  Even if you're not.
```

Here our Folk Art Festival production script is cast into one-column format. Personally, I prefer two-column, since it allows you to line up narration, music, and sound effects against the slides more precisely. But some directors like the one-column format because of the extra space it gives them for production notes. Your copy column may go on either left or right, or even down the middle. Suit your producer's preference.

```
                                        Festival--11


         INT. ; EXHIB. HALL - DAY

    71   MLS, to take in food booth area.
                             GUS
                    Hey, now!  Art or no art, folks
                    do get hungry, don't they?
                    Well, Crossroads is ready for
                    'em.  Come one, come all!  It's
                    chow time!

    72   MCU, happy-looking visitor couple.  Man's pointing off
         slide.

    73   MLS, couple from 72.  They're headed towards nearest food
         booth.

         MUSIC: Mariachi group with strongly Mexican number IN, UP
         and DOWN to BG.

    74   MCU, tamale booth.
                             GUS
                    The tamale table's great, if
                    you're a border type...Even
                    if you're not.

    75   MS/MCU, fat, happy-looking Mexican in chef's outfit--
         Diego de Vegas, maybe.  Have him dipping into tamale pot.

    76   CU, plate of steaming tamales.  They look great.

    77   MCU/CU, couple from 72.  One's taking a bite of tamale.
         The other holds a partially consumed tamale.  He/she looks
         delighted.

         MUSIC:  Combo (strong on bones and banjo) with Dixie IN, UP,
         and DOWN to BG.

    78   MS, barbecue booth...different couple.  Black woman (Delia
         Walker?) is serving.
                             GUS
                    Barbecue ribs, Crossroads
                    style.  Old South eatin'.
                    You-all don't know how good
                    ribs can be till you taste
                    these.

    79   CU, different couple at booth.  Angle favors man as he
         bites rib, sauce dribbling down chin for laughs.

         MUSIC:  Combo with Deep in the Heart of Texas IN, UP, and
         DOWN to BG.
```

Another one-column format. This one is full page width.

```
                                        Festival--11


9    INT. - EXHIB. HALL - DAY
                              GUS
              Hey, now!  Art or no art,
              folks do get hungry, don't
              they?  Well, Crossroads is
              ready for 'em.  Come one,
              come all!  It's chow time!

       Our slide takes in the entire food booth area.  Picking
       up a happy-looking visitor couple, we point them towards
       the provender...zero in on the tamale booth.  MUSIC builds
       mood with a strongly Mexican number.

                              GUS
              The tamale table's great, if
              you're a border type...Even if
              you're not.

       Our couple's at the tamale booth.  It features a fat,
       happy-looking Mexican in a chef's outfit--Diego de Vegas,
       maybe.  He serves the couple.  They eat, looking
       appropriately delighted.

       The barbecue booth.  MUSIC (strong on bones and banjo) hits
       Dixie.

                              GUS
              Barbecue ribs, Crossroads
              style.  Old South eatin'.
              You-all don't know how good
              ribs can be till you taste
              these.

       A different couple eating...black woman (Delia Walker?)
       serving.  Man's played for comedy--head thrust forward,
       sauce dribbling down his chin as he bites a rib.

       The chile booth.  MUSIC comes on strong with Deep in the
       Heart of Texas.

                              GUS
              Tex-Mex chili.  The flavor
              folks dream about clear
              'round the world.

       An enraptured visitor eating chili.  It's served by Adolph
       Fredericks.  He looks ferocious--glaring, mustache quivering,
       etc.

                              GUS
              That chili...Men have fought
              an' died to protect their
              private recipes.
```

The master scene format, as applied to our Folk Art Festival slide show. While it's fine for entertainment film, I really don't care for it in other AV situations. Use it only when you're flying blind and are in no position to spell out your shots. Why? Because this format gives the director total control. Whether he'll bring back the slides you need remains open to question. Note that you number by sequences only in the master scene format.

There are, indeed, programs that involve no narration whatever. Disney's *It's a Small World* comes instantly to mind, and so do the Chaplin films, many MTV programs, assorted dance and art presentations, and a host of silent filmstrips.

The value of such materials runs high in the right circumstances. Perhaps the most obvious case in point is that of the AV program that is to be used with audiences that don't speak the language in which the program was produced. There, the narrated module may prove next to useless, whereas the one without words makes its teaching point or creates its mood and provides its designated experience nicely.

In other words, narration can go in a variety of directions; it can be handled—or sometimes eliminated—in a host of ways. As in so many aspects of AV, there are times when you have to use your own judgment. But—back full circle to starting point—*in general*, video should dominate and tell the story.

This is just as true when you're scripting lines to be recorded on cassettes to guide visitors touring an art gallery or museum (Hampton Courts outside London, for example, or Paris's Louvre) as when you're working up a slide show. Audio doesn't just describe visuals; it comments on them. In other words, don't tell the audience what's already obvious in the visual. This is also true for button-activated commentaries at displays, exhibits, or dioramas.

What about lip-synchronized dialogue? By and large, it finds little place in AV work, save perhaps for in-house video. To a large degree, you learn the skills involved by doing. For a detailed discussion, see my *Film Scriptwriting*.

How do you write effective narration? Your starting point, it seems to me, lies in your project treatment and organization of your material. As you develop your production script from the treatment, arrange your visuals in appropriately logical order, simultaneously setting down whatever interpretive commentary—that is, narration—that seems desirable.

The next step is to check back to make sure you've included all the factual information that isn't conveyed by the visuals and *must* be presented.

Why do I emphasize "must"? Because it's essential, in narration, to hold down the length. Do your best not to let yourself be lulled by the allure of your own words into falling into a lecture approach. Don't be afraid of silence, or letting music or sound effects fill a void or create a mood. Your viewers may need time to absorb particular visuals.

Someplace along the line, you'll also need to decide on the tone you want to strike in your narration. Is it to be light, pontifical, folksy, formal, colloquial, voice-of-doom, or what have you?

Here, experimentation is your best tool. Try key lines first one way, then another, until you find an approach that sounds right. Don't be too quick to compromise. Tone can be vital!

Beyond this, as you work up and smooth and polish, strive for our old friends clarity and simplicity first of all. You'll find second person (the "you" approach), active voice ("He saw the snake" not "The snake was seen"), and the simple declarative sentence ("They moved to the next mound") will help sharpen your lines immeasurably.

Strive too for "talking writing," the effect of speech, rather than a literary style. Avoid figures and statistics as much as possible. They're deadly! Don't describe precisely what's seen; interpretive comments will make your point better. And certainly don't talk about things *not* seen, except as you interpret things that are. (We've all seen the video that mentions skiing on the nearby mountains while we're watching swimmers in the surf. It always strikes a strange—and often humorous—note.)

Anything you can do to rouse interest with narration will be appreciated. Dramatization, humor, and the use of more than one voice can be effective in their place. So, again, don't be afraid to experiment.

As you write, read your narration aloud or persuade a friend to do so while you listen. You'll be shocked at the awkward twists and stiffnesses that will be revealed.

In addition, reading aloud will help you with cueing—that is, timing out the precise moment or point within a visual at which narration should begin—and determining how long any given segment of narration should run. Although all sorts of formulas have been devised for setting narration length, your own ear is still the best judge as to how long a speech can run before boredom sets in.

## KEEPING IT SIMPLE

Back to the two-column script. With your typewriter or word processor set for pica type (10 pitch or equivalent) and the paper's left edge at zero, descriptions of visuals will ordinarily run from spaces 15 to 40; sound descriptions, 45 to 70. Page numbers will come two spaces below the top of the page; the first line, four lines below the page number. All copy should be double spaced with four spaces between shots.

How do you describe the visuals? As you'll see from the samples in Part Three, it's pretty much up to you. The big issues are clarity and simplicity. For my money, that means emphasis on pictorial nouns, action verbs, and simple declarative sentences.

I also have prejudices about this phase of format. If a presentation is to include both art and photographs, I like to have each labeled as such. The same goes for video and computer-generated graphics.

Further, I believe crews work faster and better if there's no confusion as to conditions of work. It helps to specify whether a picture is INT. (interior) or EXT. (exterior), its locale (PARK, GROCERY STORE, COTTAGE, SURGERY), and the time (DAY or NIGHT). This information should be typed ALL CAPS at the visual's start.

Below this, you should tell how large the subject is to appear in the picture, relative to the background, and what the subject is doing. Film terminology on this works from three basic positions: LS (long shot), which describes the subject in relation to background, MS (medium shot), a shot that takes in the subject but not much else, and CU (closeup), an emphasis shot that calls attention to some limited portion or specific detail of the subject.

Other common designations include ELS (extreme long shot), MLS (medium long shot), MCU (medium closeup), and ECU (extreme closeup). With all of these shots, however, the important thing to remember is that they are relative. Consequently, the image size designations are meaningless unless you name the subject: "ELS, brown cow," "CU, dagger handle," "MCU, Genevieve," or the like.

On the audio side of the page, set down anything that should be heard, whether MUSIC, SOUND EFFECTS (FX, SFX), NARRATOR (NAR.) or dialogue (ED, CORNELIA, GOD). If the music you want is "Ride of the Valkeries," obviously you must state that. This is also true for thunderbolts, Porsche motors revving, or the chirping of crickets where sound effects are concerned. The narrator and actors will appreciate it if you let them know you want them to speak wearily, angrily, hoarse with fright, or what have you.

One-column scripts use the same terminology as shown above, with visuals and sound written in the same column.

The master scene script describes action from spaces 15 to 75, gives dialogue lines or narration from 30 to 60, sets up parenthetical business (the manner in which speeches are delivered) from 40 to 55, and names the speakers from 45. Speeches, narration, and action are single spaced. Everything else is double spaced.

That is, sometimes that's the way it is. But as you'll see from our Part Three samples, this is a business full of individualists. Some will single space where I say double, some will use capitals where I recommend lower case letters, some will describe shots and settings in a manner to make my hair stand on end. No matter. You still have a guide of sorts to go by, adaptable to whatever circumstances you find yourself confronting.

And do remember always that the format your boss specifies is the correct format.

## MAKING IT COMPLEX

One of the first things you learn in any kind of writing is never to let anyone see rough or unfinished copy.

That warning goes double for AV shooting scripts.

The reason for this is that the best script you can write is still rough and unfinished where your sponsor is concerned. The sponsor thinks in terms of the completed program or presentation. The fact that it goes through a "paper" stage en route, or that this stage might conceivably be beyond the sponsor's comprehension, is an item that eludes him or her.

This is in no way calculated to question the sponsor's intelligence or good intentions. He or she may be a saint already canonized and a genius at electronics or finance or mass production to boot. Yet in all too many cases, the script *is* beyond the sponsor—so far beyond that he or she will likely never grasp it. Why? Because, like millions of others, the sponsor lacks the power to visualize and simply won't be able to make the jump from words to pictures; the sponsor won't be able to hear sound and see pictures from the pages you present.

Naturally, this isn't true of *all* sponsors. It occasionally will be your good fortune to find yourself working with one who sees your concepts every bit as vividly as you do, and maybe more so. In which case, congratulations!

More often than not, however, you won't be so lucky. In addition, the sponsor almost certainly will insist on seeing your shooting script. Then, having seen, he or she will panic.

I can offer no sure-fire solution to this dilemma. My only suggestion is, in essence, cautionary: Don't *you* panic also.

To this end, put yourself in the sponsor's place. Assume he or she is a microbiologist. Called into the laboratory, you're told to look into a microscope. Will you see the same things that the sponsor does and interpret them in an identical manner?

The answer, of course, is no. The microbiologist, as a specialist, sees and interprets with a specialist's knowledge.

You, as a specialist in scripting for the audiovisual media, are in a similar position. Looking at a script, you see things in it and interpret them differently than does—indeed, than *can*—a nonspecialist.

How do you cope with this? For one thing, don't just hand the sponsor the script. That's no more fair than it would be for the microbiologist to command you to have a go at the microscope.

Rather, make it a point to let the sponsor know in advance that script interpretation calls for its own brand of expertise. It will do no harm, indeed, if you parade a few polysyllabic technicalities, refer to Marshall McLuhan, Sergei Eisenstein, Slavko Vorkapich, and Alfred Korzybski, or throw in overtones of the esoteric.

Then, *act out* your presentation, shot by shot and step by step. Don't hesitate to dramatize, gesticulate, even leap about, if that seems called for. Radiate enthusiasm. Pulse reassurance. Make it definite to the sponsor that it's clear sailing and that there are no problems. A bit of flattery in regard to the sponsor's acumen and insight might not even be amiss.

The sponsor, in turn, will have ideas. Listen to them, separating them into two categories as you do.

Some will be generalities: overall reactions to your product.

Others—the overwhelming majority—will focus on specifics and pinpoint details.

Check the generalities carefully. It just may be that the sponsor has caught some weakness you've overlooked or a strength that rates greater attention.

Pay attention to the specifics too, for again the sponsor may have noted flaws or errors. But by and large what he or she has to offer will be nit-picking based on ignorance of the field. Typically, the sponsor will complain about your choice of words in describing visuals, not realizing that those words are intended to make things clear to the crew and so have only indirect bearing on what reaches the screen. Or the sponsor will think that a closeup means that the camera is necessarily close to the subject or that punctuation doesn't follow the rules he or she learned in school or that all sorts of miscellaneous background information needs to be incorporated into shot descriptions.

Whatever the sponsor says, work through the script with him or her step by step and shot by shot. Within the framework of this process, bow to the sponsor's every whim. Let him or her change commas and abort adverbs to his or her heart's content. You can accept this with good grace and cheerful mien because, if you've handled things skillfully, all that's sacrificed is a little time, and you know the sponsor's not going to spoil your brain child after all.

## THE STORYBOARD APPROACH

A storyboard is a script translated into still pictures. It consists of a series of photos or sketches, each showing a successive slide, frame, or action segment of the proposed AV presentation. Ordinarily, it reveals only what the audience will see on the screen, though occasionally writers include diagrams, floor plans, or the like, either for their benefit or the director's.

A special pad may be used for storyboarding. It has space for sketches down the left-hand side, while the right is reserved for narration, sound, or comment. The pictures may also be centered and narration typed or written beneath each one.

Another way of handling the job is to present each picture on a card or sheet of paper that has a ratio of approximately three units of height to four of width (3 × 4", 4 × 6", or 5 × 8" pads often are used). These pages then are pinned to a corkboard or other display surface where they can be rearranged at will if concepts change or additions or deletions are made.

The storyboard's function is to help the writer, sponsor, or their associates to see the script's visual relationships and development more clearly. Also, quite often, the writer has trouble finding language to convey to others the desired image. When that happens, a storyboard's a handy tool to have in reserve.

It's also good for presenting the idea to a board of directors or other group. Pictures are entities clients tend to understand a good bit better than words, so don't let the difficulties scare you off if a storyboard's what you need. Plan out your presentation shot by shot; then sketch the key shots, however roughly, as you see them in your mind's eye. Practice is the key, and it's a great way to pass the time during television commercials.

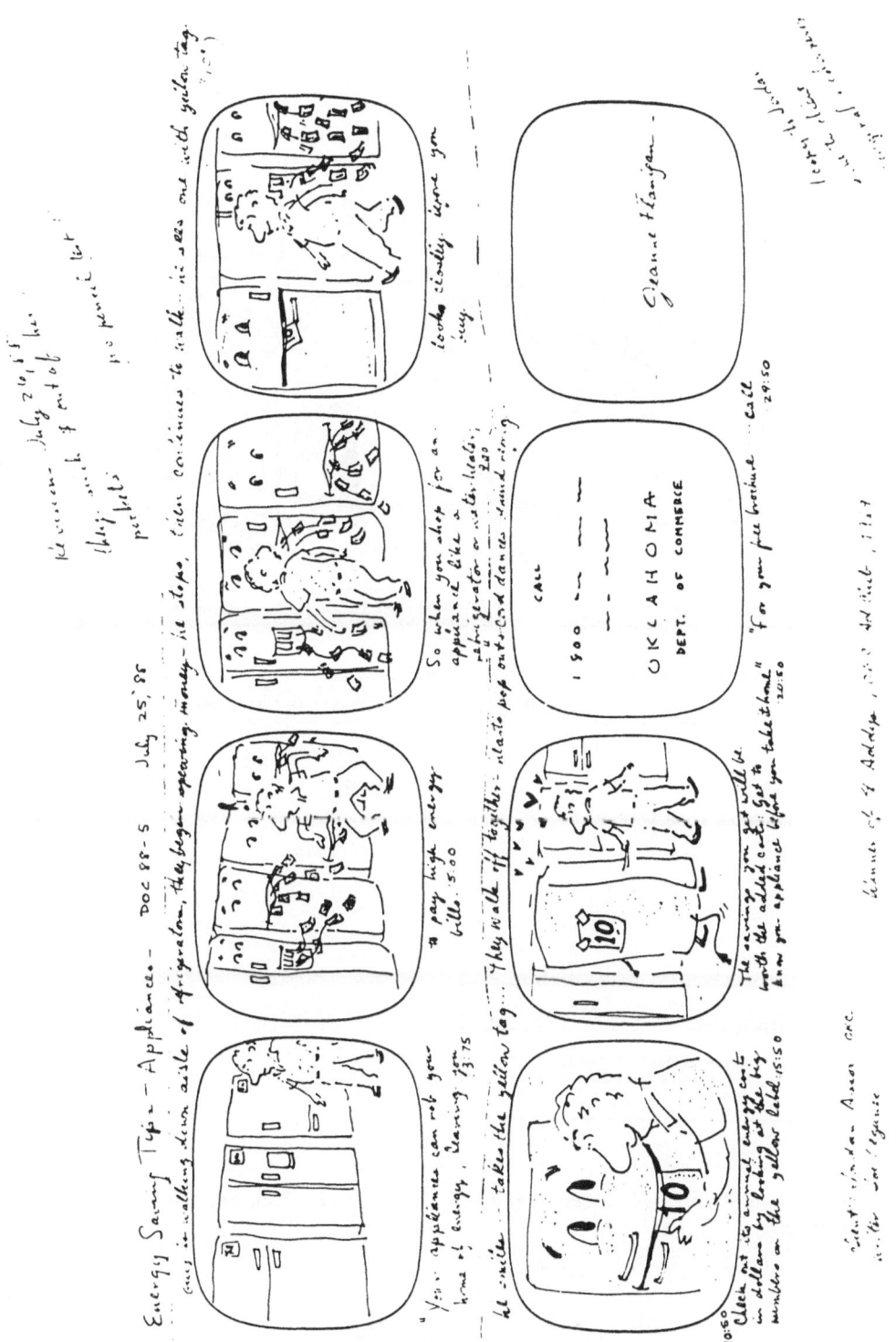

(Courtesy of Jeanne Flanigan and Jordan Associates.)

FILM: PEST CONTROL #5

FRAME NUMBER: 1

FILMSTRIP FRAME DIMENSION
35mm SLIDE DIMENSION

FINAL MEDIA - PHOTOGRAPH: ____x____  ART: _____  OTHER: _____

EDUCATIONAL OBJECTIVE: TO DEMONSTRATE PRECAUTIONS NECESSARY IN PUBLIC AREAS OUTDOORS.

VIDEO:

SHRUBBERY BEING SPRAYED BY
WORKER.  THE AREA IS ROPED
OFF.

AUDIO:

APPROVED BY: _____

(PLEASE MAKE ADDITIONAL
COMMENTS ON BACK OF FORM)

Storyboard excerpt. (Courtesy of the U.S. Postal Service Training and Development Institute.)

FILM: PEST CONTROL #5                                    FRAME NUMBER: _____

2

FILMSTRIP FRAME DIMENSION
35mm SLIDE DIMENSION

FINAL MEDIA - PHOTOGRAPH: _____ X _____   ART: _____   OTHER: _____

EDUCATIONAL OBJECTIVE: TO DEMONSTRATE THAT PUBLIC AREAS ARE TREATED AFTER HOURS IF POSSIBLE.

VIDEO:                                                   AUDIO:

WORKER SPRAYING BASEBOARDS IN
WINDOW AREA.  THE AREA IS CLOSED.
ALL WINDOWS OBVIOUSLY CLOSED.
LIGHTS ARE OFF.

APPROVED BY: _____

[PLEASE MAKE ADDITIONAL
COMMENTS ON BACK OF FORM]

FILM: PEST CONTROL #5                                          FRAME NUMBER: 3

FILMSTRIP FRAME DIMENSION

35mm SLIDE DIMENSION

FINAL MEDIA - PHOTOGRAPH: ___X___   ART: _____   OTHER: Burn/In

EDUCATIONAL OBJECTIVE: DEMONSTRATE IMPROPER PRECAUTIONS IN OPEN PUBLIC AREA.

VIDEO:                                  AUDIO:

BAIT STATION IN THE MIDDLE
OF THE FLOOR IN A POST
OFFICE BOX AREA.  B/I
LARGE "X".

APPROVED BY: _____

[PLEASE MAKE ADDITIONAL
COMMENTS ON BACK OF FORM]

FILM: _____PEST CONTROL #5_____     FRAME NUMBER: ___4___

FILMSTRIP FRAME DIMENSION
35mm SLIDE DIMENSION

FINAL MEDIA - PHOTOGRAPH: ____X____     ART: _____     OTHER: _____

EDUCATIONAL OBJECTIVE: ___DEMONSTRATE PROPER APPLICATION IN FOOD PREPARATION AREA.___

VIDEO:                                          AUDIO:

WORKER SPRAYING EMPTY CABINETS
IN FOOD PREPARATION AREA.

APPROVED BY: _____

[PLEASE MAKE ADDITIONAL
COMMENTS ON BACK OF FORM]

Forget about presentation and shooting boards. They should be the decision and responsibility of the producer or director, not the writer.

In the Postal Service art storyboard examples—they represent a "shooting" board—note how every detail is spelled out. Even the "educational objective" is nailed down, shot by shot.

Note that such a procedure is costly. In a large organization, however, it can be justified in terms of reduced shooting time later. It virtually eliminates the possibility of error in the finished product by the precision with which it shows details that could never be put across in words.

Words *are* the writer's tools, however; nowhere are they more important than in scripting.

## AGAIN, CONTINUING ELABORATION

A script is built, not born. Each time around, you flesh it out in greater detail.

Thus, from topic you go to idea.

From idea to concept.

From concept to key point.

From key point to proposal.

From proposal to treatment.

From treatment to production script.

And there it ends.

Or does it?

No. Not if the finished product's to reach its full potential.

First glances and first guesses can be good.

Sometimes. But not always.

When they aren't, you need second looks and second guesses.

Maybe even third.

It's vital, therefore, that you not try to go too fast or to freeze your thinking before it's solid. Rather, bide awhile, and brood awhile also. Make further notes. Question your shots, your approaches, your ideas.

Can it be that your pages have become too cluttered and you need to retype, reworking as you go? Then do it!

Talk things over with the producer. You might even go so far as to listen to what he or she has to say. *Really* listen.

Is it possible that some of the sponsor's notions actually aren't so crackbrained? Don't shut them out just because the sponsor's the one who voiced them.

You might want to check out your final script to be sure you actually covered the points you set out to make in the beginning. Indeed, it's worth your while at the beginning to make a list of these points so that you can consult them later and ensure you've included every one of them.

Above all, try to visualize your finished project: the program you hope ultimately to see or feel or experience. Analyze it, step by step and shot by shot and display by display. Try to spell out your goals more clearly.

You'll never reach a state of total satisfaction, of course. Nirvana is something to dream of, not to attain, and deadlines put an end to every project. But at least you'll know you've given the job your best shot. That's worth something.

## CASE STUDY 7

*Putting* Guy with an Itch *down on paper in a shooting script was pretty much routine. You already know the rules for that—sound on the left, visuals on the right, and so on. The sample pages tell the story.*

*After that there was "just" the matter of production—a huge task for the animators. And the singer, still singing, wrapped it up:*

> *Well, mankind still lives on*
> *In a black, black cave.*
> *But now we dig it,*
> *Cradle to grave.*
> *'Cause most every day,*
> *Some sharp guy with an itch*
> *Lights up a fresh spot*
> *With scratch and switch.*

*The kids, the audience? They loved it and went forth whistling (and thus remembering) its theme to the point that their teachers were driven to the verge of hysteria.*

*Oh, and one other detail. Guy was first-place winner among animation films in 1980's "Focus on Oklahoma" awards.*

*             \*         \*         \**

*Was the development of this script as neat, orderly, and step by step as this case study indicates? No, of course not. When ideas come thick and fast, they frequently tumble over each other in a manner that ignores logic. Notions on how to handle narration may hang in space long before details of content are established. Steps of development often are still being juggled almost to the typing of the final script.*

*Overall, however, this presentation pretty much spells out the way things happen. You can't go too far wrong if you follow it.*

NSTA #2--1

FADE IN:                         MUSIC:  IN AND UP.
1  Scarlet BG.                    THEME...guitar.


TITLES OVER:

     GUY WITH AN ITCH

(OTHER TITLES AS DESIRED)
              DISSOLVE TO:


2  The spark-spattered blackness
   of outer space.  Camera PANS
   or TRUCKS across the galaxies
   ...picks up Earth...ZOOMS or
   CUTS in on it to frame-filling
   CU.
              DISSOLVE TO:


3  Primeval landscape, complete
   with stylized crags, dinosaurs,
   etc.  Camera PANS slowly to
   cave entrance...ZOOMS in till   SINGER:
   the black of the cave fills     Now mankind used to live
   the frame.                      In a black, black cave;
                                   It's scary and spooky
                                   And man, who's brave?

(Courtesy of the National Science Teachers Association.)

NSTA #2--2

4   White eyes pop on **in** the dark-
    ness, pair after pair.  PAN-
    NING, camera centers on stair-
    stepped row of three pairs.

5   MCU, outline drawing of three    SOUND:  HIDEOUS MULTIPLE ROAR.
    roaring monsters, full-face.

6   CU, outline drawing of roaring   SOUND:  TENOR ROAR.
    Monster No. 1.

7   CU, outline drawing of roaring   SOUND:  BARITONE ROAR.
    Monster No. 2

8   CU, outline drawing of roaring   SOUND:  BASS ROAR
    Monster **No.** 3

9   MS-MLS, monsters' eyes as in
    shot 4.  Camera PANS to
    another cluster of eyes.

NSTA #2--13

FADE IN:                          SINGER:

83  MS, storm-clouded sky.  Camera   They're a wonderful breed,
    PANS while lightning flashes     All these guys that itch;
    pop on and off.                  The next to show
                                     Made a different pitch.

84  MLS, hero.  He runs toward the   He flipped over fire,
    shelter of a large tree.         And just how did it work?
                                     Then went knocking rocks

85  MS, hero...under branches now.   And made it perk.

86  MS, sky.  Pop on big lightning   SOUND:  THUNDER.
    bolt.

87  MLS, hero.  He stands close to
    the tree-trunk, ducked nearly
    double in panic.  Pop on bolt
    of lightning.  It hits tree.
    Pop tree to electric blue out-
    line.

88  MCU, hero.  He leaps high,       SOUND:  YELL.
    yelping and holding his shock-
    ed behind.

89  CU, fire blazing from light-
    ning-shattered tree trunk
    (hero's POV).

NSTA #2--20

132   REPEAT SHOT 18.                    SINGER:

                                         Oh, the cat that counts

133   REPEAT SHOT 19.                    Is the guy with an itch

                                         To turn on lights

134   REPEAT SHOT 13.                    With a brain-power switch.

                                         He's a curious cat,

                                         And his rat's a dog

                                         That don't dig stumbling

                                         In a fog.

                                         He's got to know,

                                         Come moan or groan,

                                         So he wears out his knuckles

                                         On his old head-bone;

                                         But he builds new ideas--

                                         Like stone by stone;

                                         Sharpens 'em with

                                         That re-search hone

                                         Till he scratches new windows

                                         In the black unknown.


135   REPEAT SHOT 14.                    SOUND:  REVVING MOTOR.


136   REPEAT SHOT 15.                    SOUND:  POING-G-G!!!


137   REPEAT SHOT 16.


138   REPEAT SHOT 24.

225  MS, junior high classroom...
all kids with features of our
hero. Camera shoots class
full face from front of room
...PANS slowly...ZOOMS IN on a
boy and girl (seated side by
side) to MCU chest two-shot.

226  CU, girl. Raptly, she stares          GIRL (VOICE OVER): Gee, I'm
off into space.                        curious sometimes. If I take
                                       the right courses and study hard,
                                       maybe when I grow up I can be a
                                       scientist too!

227  CU, boy...starry-eyed, intoxi-        BOY (VOICE OVER): Who wouldn't
cated with his own thoughts.           get a kick out of digging into
                DISSOLVE TO:           stuff that way...learning all
                                       about what makes things tick?
                                       That's for me, all the way!

228  MCU chest two-shot, boy and
girl. Camera PULLS BACK to LS
of class.
                DISSOLVE TO:

NSTA #2--37

229  MS, Earth in space. There's a   MUSIC:  IN AND UP ON THEME...
     flash of light.  A Rocket       THEN DOWN AND OUT, TIMED TO
     streaks out into space.  Cam-   FADE.
     era PANS, following action and
     holding rocket centered as it
     disappears into the kind of
     star-spattered blackness on
     which we opened.


     FINAL TITLES OVER:


               THE END


     (OTHER TITLES AS DESIRED)
               FADE OUT



                  The End



     Dwight V. Swain
     Univ. of Oklahoma
     Norman, Oklahoma

# PART THREE

# WORKING IN THE SPECIALTIES

---11---

# The Slide Show
# Today

For the purposes of this book the term *slide show* refers to any linear presentation of still images with accompanying sound.

Sometimes. As you'll see, circumstances alter cases, and terminology can be warped to take in a wide variety of things.

Traditionally, the slide show projected a series of 35mm film transparencies onto a screen, or slides were printed in sequence to form a film strip, a continuous piece of film.

Today a slide show can be recorded on videotape; it can be programmed into a computer and projected onto a screen or shown on the computer monitor. It may even be combined with action footage. So, regardless of the technology used to record the images or to project the images for audience viewing, this segment deals with linear presentations of still images.

Where sound is concerned, the simplest of so-called slide shows may have the person in charge inserting slides into a projector one by one and talking about the pictures as they flash onto the screen. More sophisticated versions use magazine-fed automatic projectors with commentary recorded and synchronized. Others may be totally automated to include stereo sound and multi-images.

The slide show is one of the commonest audiovisual forms. It can be one of the simplest to produce. Both audio and visual aspects are easily changed simply by making a new tape or substituting, inserting, or deleting a slide or a sound unit.

Scripting, however, remains vital. To make a successful slide show, you need a good script.

That means one where the writer has borne purpose and audience in mind. The writer must consider what the sponsor wants viewers to do when the lights come on and the existing attitudes of the viewers.

Consider the Ditch Witch script, for example. Clearly, it's directed toward Ditch Witch people. Its goal is to improve the company's image

CLIENT:     CMW/Ditch Witch
SUBJECT:    DWPS A V
JOB NO.:    110-9 DWPS
DATE:       Thu, Jan 25, 1990
DRAFT:      FINAL

| No. | VISUAL | AUDIO |
|---|---|---|
| 1. | Slow fade-up from black. | (Music begins.) |
| 2. | Ditch Witch service garage door in early morning light. | (SFX of early morning sounds, crickets, etc.) |
| 3. | Progressive slides of garage door being raised up. As it rises, it reveals the Ditch Witch service manager, as he lifts the door up. | (We hear an industrial-type garage door raise.) |
| 4. | Dissolve to CU of coffee being poured into DW mug. | (SFX of coffee pouring.) |
| | | (Announcer VO Male & Female voices alternate as appropriate) |
| 5. | Exterior shot of Ditch Witch dealership silhouetted in early morning light. | (M) A NEW DAY IS BEGINNING AT THE PARTS AND SERVICE DEPARTMENTS OF DITCH WITCH DEALERS ACROSS THE COUNTRY. |
| 6. | Interior shots of service manager at desk with adding machine totaling receipts for the month, leaning back in chair thinking. He looks at some old Ditch Witch promotions. | (F) YOUR PEOPLE WORK HARD SUPPORTING YOUR CUSTOMERS—AND THE PARTS AND SERVICE BUSINESS IS THERE--BUT IS THERE ANY WAY TO BUILD A CONSISTENT PARTS AND SERVICE VOLUME? |

(Courtesy of Ditch Witch and Jordan Associates.)

CLIENT:     CMW/Ditch Witch
SUBJECT:    DWPS A/V
JOB NO.:     110-9 DWPS
DATE:       Thu, Jan 25, 1990
DRAFT:      FINAL

| No. | VISUAL | AUDIO |
|---|---|---|
| 7. | Shots of Chain Challenge promotion materials and other parts promotions flyers, mailers, etc. | (M) SURE, DITCH WITCH FACTORY PROMOTIONS LIKE THE CHAIN CHALLENGE HELP BRING IN MORE PARTS AND SERVICE BUSINESS— IF ONLY THERE COULD BE MORE OF 'EM!  AND WHEN THE PROMOTIONS *DO* HIT THE STREETS, THE TIMING ISN'T ALWAYS GOOD FOR YOU... |
| 8. | Parts department manager with ruler, scissors and markers designing amateurish brochure. | (F) AND WHO'S GOT THE MONEY OR TIME TO THINK UP, PRODUCE AND MANAGE A STEADY STREAM OF PARTS AND SERVICE PROMOTIONS? ... |
| 9. | Shots of several salesmen with customers and Ditch Witch equipment. | (M) IT'S NOT AS IF YOU CAN GO OUT AND *SELL* DITCH WITCH SERVICE LIKE A NEW PIECE OF IRON ...(pause)  OR IS IT?... |
| 10. | Parts and Service sales graph showing one line with dramatic high and low points. Dissolve to same graph with sales line leveled over the course of the year. | (music transition) (F) WHAT IF YOU COULD CREATE A STEADY STREAM OF PARTS AND SERVICE BUSINESS THAT LEVELED YOUR PEAK AND SLOW SEASONS? |
| 11. | Graphics slide of Ditch Witch promo mailers streaming, domino style into infinity.  Computer graphics title: | (M) WHAT IF YOU COULD HIT YOUR CUSTOMERS AND PROSPECTS WITH A CONTINUING STREAM OF COMMUNICATIONS, SO YOUR DITCH |

| | CLIENT: | CMW/Ditch Witch |
|---|---|---|
| | SUBJECT: | DWPS A/V |
| | JOB NO.: | 110-9 DWPS |
| | DATE: | Thu, Jan 25, 1990 |
| | DRAFT: | FINAL |

| No. | VISUAL | AUDIO |
|---|---|---|
| | Parts & Service Promotions | WITCH BUSINESS WOULD BE ON THEIR MINDS MORE OFTEN? |
| 12. | Split screen image of many customer faces. Ditch Witch logo dissolves over customer faces. | (F) WHAT IF YOU COULD CREATE FOR THE DEALERSHIP A STRONG IMAGE OF PROFESSIONALISM AND QUALITY IN THE MINDS OF YOUR CUSTOMERS, EVEN WHEN THEY'VE NEVER SEEN THE DEALERSHIP? |
| 13. | Customer stands at DW parts counter. A parts department person is handing him a replacement part. Both people are talking and smiling. They obviously know each other. | (M) WHAT IF YOU COULD CREATE A STRONGER SENSE OF LOYALTY IN YOUR CUSTOMERS SO THEY KEEP COMING BACK TO YOU FOR DITCH WITCH PARTS AND SERVICE. |
| 14. | Sequential shots of calendar as the months flip past, from January to December. | (F) WHAT IF YOU COULD DESIGN A YEAR-LONG PARTS AND SERVICE PROMOTION PROGRAM THAT PRACTICALLY RAN ITSELF? |
| 15. | Over-the-shoulder shots of service manager with hands on hips, watching his employees working. The service garage is filled with machines and activity. | (M) THAT BROUGHT IN A REGULAR FLOW OF TRAFFIC THROUGH YOUR SERVICE AND PARTS DEPARTMENT. (SFX garage repair sounds, not much activity. As slides dissolve the SFX changes to the sounds of a service garage bustling with work and activity.) |

3

*Jordan Associates*

and to create a sense of loyalty in their customers. The key issues of audience and purpose are established early.

Not every script is this direct. In some, well-nigh absolute control is demanded, in order that a particular emotional effect may be achieved.

To that end, the show will quite possibly eliminate reality completely in favor of a "story" approach—a totally fictionized and dramatized *simulation* of events.

As an example, consider "Candles on the Cake," a slide show designed to sell retirement annuities that I scripted for Rocket Pictures in Los Angeles some years ago. It was intended for use by salespeople with table-top rear-screen projectors. The idea was to persuade the prospect to grant the salesperson 15 or 20 minutes of uninterrupted time in the prospect's own office. There the salesperson would set up the automatic slide-plus-sound projector. For the allotted period, the prospect was transported into the world the show created, with no telephone or secretarial interference. When the presentation ended, the salesperson moved in for the kill, adapting the annuity idea to the prospect's personal situation and needs.

I wrote the script to appeal to the deep-rooted, nearly compulsive hunger for security and, obversely, the fear of insecurity, that lurk in so many of us.

To maximize its appeal, the script was developed within a predetermined pattern of proven effectiveness. Every video frame and audio line then was scrutinized by technical advisors and staff. Endlessly, it seemed, pictures and narration were changed and juggled to sharpen the effect and drive home the key point.

What *is* the key point? That with forethought and proper planning—that is, the purchase of retirement annuities—the prospect may anticipate a happy and secure old age. Without it, he or she may very well fall victim to disaster.

The pattern, in turn, develops this theme in terms of the life of one man. His story is told within the framework of a show of about 100 shots, with every bit as careful attention to detail as any major Hollywood production.

(For a consideration of the principles and techniques of dramatic writing in more detail, see my *Techniques of the Selling Writer* and *Film Scriptwriting: A Practical Manual*, Second Edition.)

The first half of the presentation, 62 shots, is a sort of burgeoning horror tale in which, after opening bliss, everything goes wrong. The second half replays the man's final days in happier form, on the premise of his having invested in a retirement annuity at the proper time. All of which adds up to a combination of dramatic approach and emotional impact that's hard to resist.

There are other variations on slide shows, including the use of multiscreen presentations (several projectors, sometimes with several screens). In essence, this calls for little that's different from the single-projector,

Candles--1

| 1 | FOCUS FRAME | (SILENT) |
|---|---|---|

2   ART: TITLE                 (MUSIC: INTRO)
    "B.S.B. PRESENTS..."

3   ART: MAIN TITLE            (MUSIC: UP)
    "CANDLES ON THE CAKE"
                               (MUSIC: TO B.G.)

4   ART: ILLUSTRATION          NAR: This is a story of birthdays
    CU, BIRTHDAY CAKE WITH     --the birthdays of
    CANDLES BURNING BRIGHTLY

5   PHOTO: INT
    CU, JOHN IN LIVING ROOM,   John Horton, chief clerk at Acme.
    LAUGHING DELIGHTEDLY AT    Today he's 65.  So,
    MATCHING CAKE ON TABLE
    BEFORE HIM

6   PHOTO: (CONT.)             the Hortons are having a little
    MS, HORTON FAMILY
    GROUPED ABOUT TABLE.       family celebration.  For this is
    SHOT FAVORS JOHN.
    ALSO INCLUDED ARE          a turning point for John; more
    ETHEL, BOB, DOROTHY,
    AND TWO CHILDREN           than just an ordinary birthday.

7   PHOTO: INT.                Acme gave him his first formal word
    MCU, JOHN AT DESK IN
    OFFICE, READING LETTER.    on the subject: official notice
    HIS EXPRESSION IS ONE
    OF PLEASURE--MAYBE EVEN    that he's reached mandatory
    REPRESSED TRIUMPH AND
    EXCITEMENT.  A             retirement age.  It's a moment
    DISTINGUISHED-LOOKING
    MAN STANDS BESIDE HIM      John's been looking forward to--

*Candles on the Cake* script. (Used by permission of Dick Weston and Rocket Pictures, Inc.)

Candles--2

8   PHOTO: (CONT.)           --a chance to relax and enjoy life.
    CAMERA DRAWS BACK TO
    MS. MAN PATS JOHN ON     Now, thanks to the company rule,
    BACK.  BOTH ARE
    BEAMING                  he can do it.  He's even relieved

                             of responsibility for making the

                             decision to quit.


9   PHOTO: (CONT.)           That's the way it is in business.
    CU. JOHN'S BEAMING
    FACE                     and he's glad.  Business retires

                             you; you don't retire from

                             business.


10  PHOTO: (CONT.)           The watch they gave him at the
    MS, OFFICE, DIFFERENT
    ANGLE.  JOHN IS CENTER   plant today--as John sees it, it's
    OF GROUP OF FOUR OR
    FIVE MEN, WITH           a symbol of success and of
    SECRETARY IN BACKGROUND.
    THE DISTINGUISHED MAN    fulfillment...recognition that the
    IS INCLUDED.  ONE MAN
    IS GIVING JOHN AN        company has observed and
    OPENED CASE WHICH
    CONTAINS A WATCH.        appreciated all the years of hard
    ANOTHER HOLDS A
    SCROLL-TYPE PAPER, AS    work he's put in.
    IF READING A
    PROCLAMATION


11  PHOTO: INT               That makes John feel mighty good.
    MCU, SHOT 6 PARTY,
    JOHN GESTURING AND       --In terms of the future, even
    SPEAKING, CAKE STILL
    BEFORE HIM.  OTHER       more so.  Because it looks like a
    MEMBERS OF FAMILY
    (THOUGH PROBABLY NOT     clear track ahead.
    ALL) ARE INCLUDED

Candles--3

12  PHOTO: (CONT.)      A man couldn't ask for a better
    CU, ETHEL, LAUGHING
                                      wife than Ethel.

13  PHOTO: (CONT.)      Son Bob is a man any father would
    CU, BOB, LAUGHING
                                        be proud of.

14  PHOTO: (CONT.)      Dorothy's a fine daughter-in-law;
    CU THREE-SHOT, DOROTHY
    WITH HER ARMS ABOUT THE and the grandchildren--well!
    TWO CHILDREN.  ALL ARE
    LAUGHING.

15  PHOTO: INT.          Through the years, John's liked
    MCU THREE-SHOT, JOHN
    CONFERRING WITH TWO   his work and the people he's
    OTHER MEN IN SHOT 9
    OFFICE.  ALL LOOK      worked with.
    GOOD-NATURED AND
    FRIENDLY

16  PHOTO: INT.          His terms on the church board--
    MS, CONFERENCE ROOM,
    CLEARLY IDENTIFIABLE AS they prove he's responsible and
    IN CHURCH.  SEVERAL
    MEN ARE SITTING ABOUT respected.
    TABLE--ONE (NOT JOHN)
    AT HEAD, OTHERS ALONG
    SIDES.

17  PHOTO: EXT           His neighbors are all his good
    MCU TWO-SHOT, JOHN IN
    YARD, LEANING ON RAKE  friends.
    AND TALKING ACROSS
    HEDGE TO NEIGHBOR.

Candles--4

18   PHOTO: (CONT.)              Sure, he and Ethel have had their
     MS, THAT PORTION OF
     YARD IN WHICH JOHN AND      share of trouble--
     NEIGHBOR STAND.  ETHEL
     KNEELS SOMEWHAT TO ONE
     SIDE OF THE TWO MEN,
     PRUNING ROSES OR THE
     LIKE, AND JOHN IS
     GESTURING IN HER
     DIRECTION.  THIS SHOT
     ALSO SHOULD INCLUDE
     ENOUGH OF THE HORTON
     HOME TO ESTABLISH IT AS
     A COMFORTABLE, WELL-CARED
     FOR, MIDDLE-BRACKET
     PLACE

19   PHOTO: INT.                 --the period when Ethel was in
     MS, BEDROOM.  DOCTOR
     SITS BY BED, READING        such poor health...
     THERMOMETER, MEDICAL
     CASE ON STAND BESIDE
     HIM.  ETHEL LIES IN THE
     BED, SO PLACED AS TO
     CHEAT ON HER AGE.  JOHN
     (HAIR DARK) STANDS TO
     ONE SIDE, BACK TO CAMERA

20   PHOTO: EXT.                 ... the long layoff during the
     MS, JOHN STANDING AT
     FACTORY GATE, BACK TO       recession...
     CAMERA.  HE'S WEARING
     HAT, COAT COLLAR TURNED
     UP AS IF AGAINST THE
     COLD.  HE IS READING
     PROMINENT SIGN: "NO
     WORK TILL FURTHER
     NOTICE"
     (or)
     ART: ILLUSTRATION
     NEWSPAPER WITH HEADLINE:
     "ACME CLOSES; 3000
     JOBLESS" OR THE LIKE

21   PHOTO: EXT.                 ...Son Bob's months as a prisoner
     MLS, BOB, SHOT THROUGH
     HIGH BARBED WIRE FENCE.     of war...
     AREA IS BLEAK.  BOB
     WEARS SHABBY ARMY UNIFORM.
     HE STANDS SLUMPED, HANDS
     IN POCKETS

Candles--5

22   PHOTO: INT.                    ...the years Ethel's mother stayed
     MCU, FRAGILE ELDERLY
     WOMAN IN WHEELCHAIR IN         with them before she died.
     BEDROOM

23   PHOTO:                         But troubles pass, John knows.
     CLOSE MS, PARTY GROUP
     AS IN SHOT 6.  SHOT            So he's worked hard and played
     FAVORS JOHN.  HE STILL
     LOOKS HAPPY ENOUGH, YET        his breaks straight...brought
     THERE'S A SOBER,
     THOUGHTFUL NOTE BEHIND         home his paycheck each and every
     HIS PLEASURE
                                    week.

24   PHOTO: (CONT.)                 Now, the house is free and clear;
     MS, ROOM.  PARTY IS
     STILL IN PROGRESS, BUT         all paid for.
     JOHN HAS RISEN AND IS
     LOOKING AWAY FROM THE
     PEOPLE, ADMIRING THE
     ROOM ITSELF WITH HIS
     HEAD COCKED SLIGHTLY

25   PHOTO: (CONT.)                 The furnishings--well, a man gets
     MS, LIVING ROOM CORNER.
     JOHN IS TAKING PIPE            used to familiar things.  Comfort's
     FROM RACK ON END TABLE
     THAT STANDS BESIDE A           more important than a lot of fancy
     COMFORTABLE-LOOKING OLD
     OVERSTUFFED VELOUR CHAIR       doodads.
     WITH FOOTSTOOL.  A
     LIGHTED FLOOR LAMP,
     VINTAGE 1925, FLANKS
     THE CHAIR.  A SMALL
     SHEAF OF PAPERS AND
     LETTERS, AND A FEW BOOKS,
     ARE ON THE END TABLE
     NEXT TO THE PIPE RACK.
     BANK BOOK IS ON TOP OF
     HEAP

26   PHOTO: (CONT.)                 There's even a little money in
     MCU, JOHN SEATED IN
     CHAIR FROM SHOT 25.            bonds and in the bank.  So,
     HE'S PUFFING AT HIS
     PIPE.  HE HOLDS BANK
     BOOK IN HIS HANDS,
     STUDYING IT.  HE LOOKS
     COMFORTABLE AND WELL
     SATISFIED.  SHOT FAVORS
     BANK BOOK

Candles--6

27  PHOTO: (CONT.)          John figures the years ahead are
    CLOSE MS, JOHN AND
    PART OF PARTY GROUP.    going to be <u>good</u> years.  Because
    ETHEL(AND PERHAPS
    OTHERS) HAS RISEN AND   all he and Ethel want from
    STANDS FAIRLY CLOSE TO
    JOHN.   SHOT STRONGLY   retirement, really,
    FAVORS JOHN AND ETHEL
    AS THEY GAZE FONDLY AT
    EACH OTHER, SMILING


28  PHOTO: (CONT.)          is independence.  They plan to
    CU, JOHN GAZING UP
    THROUGH HIS PIPE SMOKE  settle down to the pattern they've
    INTO THE WORLD OF HIS
    DREAMS                  chosen...enjoying the simple

                            things--


29  PHOTO: INT             --sleeping late...
    MCU TWO-SHOT.  JOHN
    LIES ABED, STRETCHING
    SLEEPILY.  ETHEL STANDS
    BESIDE HIM, ONE HAND ON
    HIP, LAUGHING AND SHAKING
    A BIG KITCHEN SPOON AT
    HIM AS IF DEMANDING THAT
    HE GET UP


30  PHOTO: EXT.            ...fishing...
    MS, JOHN AND ANOTHER
    MAN OF SIMILAR AGE IN
    ROWBOAT.  JOHN IS
    CASTING


31  PHOTO: EXT.            ...gardening...
    MCU, ETHEL CUTTING
    BOUQUET OF ROSES


32  PHOTO: EXT (STOCK SHOT) ...taking an occasional scenic
    MS-MLS, GRAND CANYON,
    NEW YORK CITY SKYLINE,  trip...
    OR THE LIKE

33  PHOTO: EXT.                  ...spoiling their grandkids.
    MS, AMUSEMENT PARK RIDE
    TICKET STAND AREA.          --Only then comes a rude
    JOHN'S BUYING TICKETS
    WHILE THE GRANDKIDS         awakening--
    EAT COTTON CANDY.
    ETHEL HOLDS LITTLE
    GIRL'S HAND

34  PHOTO: INT.                 --the day when John suddenly
    MCU TWO-SHOT, JOHN AND
    TV REPAIRMAN IN TV          realizes that he's drawn his
    REPAIR SHOP.  MAN HOLDS
    PICTURE TUBE AS IF HE       last paycheck.
    WERE EXPLAINING
    SOMETHING.  JOHN HOLDS
    CHECKBOOK IN ONE HAND,
    PEN IN OTHER.  HE'S
    FROWNING DOWN AT THE
    BOOK'S ENTRIES; PAYING
    NO HEED TO THE
    REPAIRMAN.  SHOT FAVORS
    JOHN

35  PHOTO: INT.                 Nor does it help when the Hortons
    MCU TWO-SHOT, JOHN AND
    ETHEL BY FRONT DOOR.        learn that social security and
    ETHEL HOLDS A SHEAF
    OF MAIL.  JOHN HOLDS        John's pension, both together,
    TWO CHECKS IN ONE HAND,
    THEIR ENVELOPES IN THE      simply won't be adequate for them
    OTHER.  BOTH HORTONS
    LOOK A TRIFLE UPSET AS      to live on.
    THEY GAZE AT THE CHECKS

36  JOHN AND ETHEL LOOKING      It dawns on them, then, that
    AT EACH OTHER.  THEY
    REGISTER WORRY, FRIGHT      they've made their retirement

                               plans well enough.  But they've

                               forgotten to plan for a steady,

                               continuing income--the income

                               that anyone, retired or not, must

                               have, in order to meet the bills

                               that go with living.

Candles--8

| | | |
|---|---|---|
| 37 | PHOTO: INT.<br>MCU, JOHN TAKING BONDS<br>FROM BOX IN BANK'S<br>SAFETY DEPOSIT VAULT | Economizing isn't enough.  Before<br>John knows it, he's dipping into<br>savings, cashing bonds. |
| 38 | PHOTO: INT.<br>MS,   BIRTHDAY PARTY.<br>REDRESS TO DIFFERENTIATE<br>FROM SHOT 6 | Another year, another birthday<br>party.  More and more often now,<br>extras seem to come up-- |
| 39 | PHOTO: INT.<br>MCU TWO-SHOT, JOHN AND<br>MECHANIC IN GARAGE.<br>THEY STAND BESIDE<br>WORKBENCH.MECHANIC<br>HOLDS TOOL | MECHANIC:  Sorry, Mr. Horton.  It's<br>going to have to be a valve job<br>this time.  These older cars-- |
| 40 | PHOTO: INT.<br>MCU TWO-SHOT,<br>OPTOMETRIST SEATED AT<br>FITTING TABLE IN<br>OPTOMETRIST'S OFFICE.<br>OPTOMETRIST HOLDS<br>GLASSES | OPTOMETRIST: Yes, Mrs Horton,<br>you definitely do need new glasses.<br>I'm surprised your headaches aren't<br>worse than they are. |
| 41 | PHOTO: INT.<br>MCU TWO-SHOT. JOHN AND<br>DENTIST STANDING BESIDE<br>DENTAL CHAIR.  DENTIST<br>HOLDS AND TAPS PLATE. | DENTIST:  A plate that's cracked<br>this badly can't be repaired, Mr.<br>Horton.  You'll have to have a<br>new one. |
| 42 | PHOTO: EXT.<br>MCU TWO-SHOT, JOHN AND<br>ROOFER IN YARD, HORTON<br>HOUSE IN BACKGROUND.<br>ROOFER HOLDS PAD<br>AND PENCIL | ROOFER:  That's the best price I<br>can make you, Mr. Horton.  Putting<br>a new roof on a house just plain<br>costs money. |

Candles--9

43   PHOTO: EXT.                        NAR.: John's first thought is to
     MS, JOHN CHECKING
     CLASSIFIED ADS IN                 find another job.
     FOLDED NEWSPAPER AT
     OFFICE BUILDING
     ENTRANCE

44   PHOTO: INT.                        PERSONNEL MAN: I'm sorry, Mr.
     MCU TWO-SHOT, JOHN AND
     PERSONNEL MAN. MAN SITS Horton.    But we need someone a
     BEHIND DESK WHICH BEARS
     SIGN "PERSONNEL MANAGER" bit younger for this position.
     JOHN STANDS BEFORE HIM,
     HAT IN HAND.

45   PHOTO: INT.                        NAR.:   Worse, John learns that
     MS, JOHN TALKING TO
     CLERK ACROSS COUNTER.              any work that pays more than
     HE LOOKS DEPRESSED.
     SIGN PROMINENT IN                  $100 a month, $1200 a year, cuts
     BACKGROUND SAYS THIS
     IS "SOCIAL SECURITY                down on his social security
     LOCAL OFFICE" OR THE
     LIKE                               benefits.

46   PHOTO: EXT.                        The car goes--John has to raise
     MS TWO-SHOT, MAN
     HANDING JOHN CHECK.                cash somewhere.
     THEY STAND BEFORE A
     BIG USED AUTO LOT SIGN
     "CASH FOR YOUR CAR" OR
     THE LIKE

47   PHOTO: EXT.                        Church donations?  Entertaining?
     MLS, JOHN AND ETHEL
     PASSING TOY STORE                  Presents for the grandchildren?
     WINDOW DISPLAY.  THEY
     WALK WITH HEADS DOWN,              No more.  Pride takes a beating
     SHOULDERS SLUMPED,
     NOT LOOKING AT THE                 when financial security fades.
     DISPLAY

Candles--10

48   PHOTO: INT.                    John's next birthday isn't quite
     MCU, BIRTHDAY PARTY
     FAVORING JOHN.  HE             so happy.  If he and Ethel didn't
     LOOKS DEPRESSED AND
     HARRIED DESPITE HIS            own their home, they might be in
     CANDLE SPARKLING CAKE
                                    real trouble.

49   PHOTO: INT.                    John's health begins to fail.
     MCU TWO-SHOT, DOCTOR
     TAKING JOHN'S BLOOD            Bills pile up.  Even with Bob and
     PRESSURE AGAINST
     MEDICAL OFFICE                 Dorothy helping, the Hortons'
     BACKGROUND
                                    pension money won't go around.

50   PHOTO: EXT.                    The Hortons' house costs more to
     MS, HOUSE.  JOHN AND
     ETHEL STAND NEAR THE           keep up than they can afford.  But
     ENTRANCE, JOHN IN
     SHIRTSLEEVES AND ETHEL         it's hard to let a place go when
     IN HOUSEDRESS.  THEY
     LOOK UPSET.  A MIDDLE-         you've 30 years of memories in it.
     AGED COUPLE IN STREET
     ATTIRE STAND OFF TO
     ONE SIDE, SURVEYING
     THE HOUSE CRITICALLY.
     WITH THEM IS A REALTOR-
     TYPE MAN IN BUSINESS
     SUIT, SPEAKING AND
     GESTURING.

51   PHOTO: INT                     Now, though, there's no choice.
     MCU TWO-SHOT, CRASS-
     LOOKING MAN AND WOMAN          Worst of all is the auction--
     WITH NEEDLEPOINT CHAIR
     OR THE LIKE TURNED             cherished possessions pawed over,
     UPSIDE DOWN SO THEY
     CAN INSPECT THE JOINTS         sold to strangers.

Candles--11

52  PHOTO: INT.                    The watch Acme gave John when he
    MCU TWO-SHOT, JOHN
    AND PAWNBROKER AGAINST         retired goes to the pawnshop.
    PAWNBROKER BACKGROUND.
    JOHN IS PUSHING WATCH          Time's run out for him, and he
    ACROSS COUNTER AS
    PAWNBROKER EXTENDS             knows it.
    CASH

53  PHOTO: INT.                    SON: Dad, I'm sorry.  But Dorothy
    MCU THREE-SHOT, JOHN,
    ETHEL AND BOB.                 and I won't have you living this
    BACKGROUND IS LIVING
    ROOM OF EXTREMELY              way.  You'll have to move in with
    SLEAZY-LOOKING
    APARTMENT                      us.

54  PHOTO: INT.                    NAR.: The birthday that follows
    MS, BIRTHDAY PARTY SET
    IN CONTEMPORARY-STYLE          the move is a sort of private
    LIVING ROOM OF BOB'S
    HOUSE.  ETHEL IS               nightmare for John.
    APPROACHING TABLE,
    CARRYING LIGHTED CAKE.
    DOROTHY IS COMING CLOSE
    BEHIND AND TO ONE SIDE
    OF HER, CARRYING TRAY
    WITH DISHES OF ICE
    CREAM.  THE GRANDKIDS
    STAND IN DOORWAY TO ONE
    SIDE.  JOHN AND BOB ARE
    ALREADY SEATED AT THE
    TABLE.  SHOT FAVORS JOHN.
    HE LOOKS EXTREMELY
    DEPRESSED AS HE STARES
    AT THE CAKE

55  PHOTO: (CONT.)                 All he can think about is how he's
    MCU THREE-SHOT, JOHN,
    ETHEL AND DOROTHY.             failed the one he loves the most,
    JOHN WATCHES ETHEL AS
    SHE TALKS TO DOROTHY.          even though she'll never say a word
    THERE'S AN AIR OF
    RESIGNATION ABOUT HER.         about it.
    SHOT FAVORS ETHEL

Candles--12

56  PHOTO: (CONT.)          Crowding both grandchildren into
    MCU TWO-SHOT, THE
    GRANDKIDS.  THEY'RE     one bedroom--well, John knows it
    STARING SULLENLY OUT
    OF FRAME               can't help but change the way

                          they feel about him and Ethel.


57  PHOTO: (CONT.)          Dorothy's a wonderful daughter-in-
    MCU TWO-SHOT, ETHEL
    AND DOROTHY.  ETHEL    law.  But what young woman wants
    IS LEANING FORWARD TO
    SET DOWN THE CAKE.      her husband's mother telling her
    DOROTHY IS LOOKING AT
    ETHEL (WHO CAN'T SEE    how to run her kitchen?
    HER FACE) WITH JUST
    THE SLIGHTEST EXPRESSION
    OF IRRITATION.  SHOT
    FAVORS DOROTHY


58  PHOTO: (CONT.)          John sees the friction it creates
    MCU TWO-SHOT, DOROTHY
    AND BOB.  THEY'RE       between her and Bob.
    STARING AT EACH OTHER
    WITH CLEARLY
    ANTAGONISTIC EXPRESSIONS


59  PHOTO: INT              In the days that follow, the kids
    MCU THREE-SHOT, DOROTHY
    AND THE GRANDKIDS, IN   grow more rebellious.
    KITCHEN.  DOROTHY IS
    SCOLDING ONE OF THE
    CHILDREN, AND THE
    CHILD'S SNARLING BACK.
    THE OTHER YOUNGSTER
    STANDS SULLENLY IN THE
    BACKGROUND.


60  PHOTO: INT              Noise, confusion, humiliation--
    MS, LIVING ROOM.  THE
    KIDS ARE RACING THROUGH John hates every aching moment of
    OR FIGHTING.  JOHN
    STANDS IN ONE CORNER,   it!  Is this the independence in
    HOLDING HIS EARS.  HE
    LOOKS MISERABLE.        retirement that he dreamed of?

Candles--13

61  PHOTO: INT                    Another year...another birthday
    CU, JOHN STARING AT
    WALL CALENDAR.  NO           coming up.  John dreads it.  No
    YEAR SHOWS, BUT THE
    DAY OF THE MONTH IS          party for him, this time; no
    CIRCLED
                                 candles on his birthday cake.

62  PHOTO: INT.                  Too late, he's learned a bitter
    MS, BEDROOM.  JOHN
    SITS ON THE EDGE OF          truth: It's just as bad for a man
    THE BED, HEAD IN
    HANDS, ELBOWS ON             to live too long as it is for him
    KNEES, A PICTURE OF
    DEFEAT                       to die too soon.  --But why must

                                 a man live till he's past 65 to

                                 find it out?

                      (END OF PART ONE)

63  PHOTO: INT.                  Yes, that's the way it usually
    REPEAT SHOT 28
                                 happens, when a man doesn't plan

                                 for retirement income.  But for

                                 John Horton, fortunately, it came

                                 to pass only in his imagination;

64  PHOTO: (CONT.)               a fantasy at his sixty-fifth
    CLOSE MS, PARTY GROUP.
    ALL ARE LAUGHING AS BOB      birthday party.  The reason dates
    AND DOROTHY PULL JOHN
    UP FROM HIS CHAIR            back some years, to John's

                                 memories of old friends--

Candles--14

65  PHOTO: INT.             --like Sam Reddick.  Sam couldn't
     MS, PRODUCE STORE,
     ELDERLY SAM REDDICK     manage on social security, after
     SWEEPING IN FOREGROUND.
     HE LOOKS PITIFUL AND     he retired.  So he ended up as
     INADEQUATE
                           sweeper at a produce store.

66  PHOTO: INT.             Ed Roberts' girls put <u>him</u> in a
     MS, LARGE BLEAK-
     LOOKING ROOM WITH      rest home, just to get him out
     SEVERAL ELDERLY PEOPLE.
     ED ROBERTS STANDS      from under foot.
     LOOKING OUT WINDOW IN
     FOREGROUND, FRAGILE
     AND FORLORN

67  PHOTO:EXT.            Lillian Goss's church circle made
     MS, LILLIAN'S FRONT
     DOOR.  THREE DO-GOODER  her their private project after
     TYPE WOMEN STAND JUST
     OUTSIDE WITH COVERED    she lost all her money on those
     DISHES.  LILLIAN IS IN
     THE OPEN DOORWAY,     bad investments.
     GREETING THEM WANLY

68  PHOTO: INT.             John saw that a man who wants
     MS, JOHN (AGE 45) IN
     SILHOUETTE AT DUSK AT   independence needs to plan ahead
     WINDOW, PONDERING, HIS
     BACK TO CAMERA       for it.  For earned income stops
                           when he retires.

69  ART: CHART           Could he <u>save</u> enough to support
     LIFE EXPECTANCY AT 60,
     65, 70 AND SO ON     him and Ethel for the rest of
                           their lives?  The life expectancy
                           tables said no.

Candles--15

70  ART: ILLUSTRATION          Investments?  For an income of
    BALANCE-TYPE SCALE
    WITH ONE PAN MARKED        only $100 a month, John would
    "$40,000", THE OTHER
    $100                       have to accumulate thirty or

                               forty thousand dollars!


71  ART: ILLUSTRATION          To get higher interest, bigger
    FRAME-FILLING DISPLAY
    OF STOCK CERTIFICATES,     returns, he'd have to gamble on
    FEATURING OFF-BEAT
    ITEMS                      speculative ventures--and maybe

                               lose his shirt


72  PHOTO: INT.                He began to wonder if there <u>was</u>
    CU, JOHN IN BED, NIGHT
    LIGHTING.  HE LIES ON      any such thing as retirement
    HIS BACK, BACK OF
    WRIST ACROSS EYES          security, where the average man

                               was concerned.


73  PHOTO: INT.                Then, one day, his insurance agent
    MS, OFFICE, AGENT AND
    JOHN SHAKING HANDS         came by to talk with him about
    ACROSS DESK.  CAMERA
    IS BEHIND AND TO ONE       retirement income annuities.
    SIDE OF JOHN, FAVORING
    AGENT STRONGLY


74  PHOTO: (CONT.)             AGENT:  In effect, Mr. Horton,
    MCU, AGENT SPEAKING.
    BOTH MEN NOW ARE SEATED    you'll be buying financial

                               security on the installment plan!

Candles--16

75  ART: ILLUSTRATION          NAR.:  Under the plan John chose,
    BULGING MONEY BAG
    MARKED "$30,000"           when he retired he'd receive either

                               a lump sum payment of $30,000 on

                               a predetermined date,

76  ART: ILLUSTRATION          or a regular monthly income for
    FAN OF MONTHLY
    CHECKS                     the rest of his life.  In either

                               case, it would be ample to

                               supplement social security.

77  ART:  ILLUSTRATION         The sum could equally well have
    MCU, JOHN'S HEAD AT
    AGE 40 IN SILHOUETTE.      been $15,000, $25,000, $50,000,
    SURROUNDING IT, ALL
    ANGLES, ARE QUESTION       or more.  The decision was his,
    MARKS OF ASSORTED
    SHAPES AND SIZES, AND      based on his own needs and
    SUCH LUMP SUM FIGURES
    AS "$25,000", "$35,000",   resources.
    "$50,000"

78  ART: ILLUSTRATION          To retire before sixty-five, all
    MONTAGE OF VIGNETTES OF
    IDEALIZED RETIREMENT       he had to do was state in advance
    ACTIVITIES--MAN PLAYING
    GOLF, LUXURY LINER         the age at which he wanted to
    ENTERING TROPIC HARBOR,
    COUPLE BENEATH COLORFUL    receive the money.
    UMBRELLA ON BEACH,
    COUPLE IN EVENING GARB
    ENTERING THEATRE, ETC.

79  PROMOTIONAL PAMPHLET       Because he invested in the annuity
    ON DESK.  HEADING SHOULD
    READ "RETIREMENT           at an early age, the premiums
    ANNUITY PREMIUM TABLE"
    OR THE LIKE                were low.

Candles--17

80  ART: ILLUSTRATION           Had John died before retirement
    SILHOUETTED FIGURES
    OF ETHEL (AGE 40) AND       age, Ethel and Bob would have
    BOY (AGE 12) STANDING
    IN FRONT OF STYLIZED        received the full amount of his
    SKETCH OF HORTON HOME.
    ETHEL HAS HER ARM           policy--income to replace the
    AROUND THE BOY.  A
    CIRCLE OF GOLD LINKS        money he'd have earned.
    COMPLETELY SURROUNDS
    HOUSE, BOY AND ETHEL

81  ART: ILLUSTRATION           Best of all, annuity income was
    CU, PAPER HEADED
    "ANNUITY CHECKLIST"  A      free and clear; non-taxable.  It
    LIST OF FOUR ITEMS
    FOLLOWS: "SECURITY?",       wouldn't cut down on social security
    "RETURNS?", "NON-
    TAXABLE?", "SOCIAL          benefits.
    SECURITY?"  AFTER EACH
    IS A HEAVY, RUBBER-
    STAMPED "OK"

82  ART: ILLUSTRATION           When John retired, he decided to
    CU, ANNUITY CHECK
    MADE OUT TO JOHN HORTON     take his annuity in regular income

                                form, so that he'd have a check

                                coming in each and every month,

                                even if he lived to be 99!

83  PHOTO: INT.                 Because he'd planned ahead, John
    REPEAT SHOT
                                could afford to be happy at his

                                party.  He knew what his financial

                                future held in store for him!

84  PHOTO: INT.                 Today, social security and the
    MCU, JOHN AND ETHEL
    WITH SUPERMARKET CART.      company pension plan take care of
    JOHN'S AT THE HELM,
    ETHEL TAKING SOME ITEM      the Hortons' basic necessities
    FROM SHELF
                                well enough.

Candles--18

85  PHOTO: EXT.                     But it's their annuity check that
    MCU TWO-SHOT, JOHN
    AND ETHEL BESIDE THEIR          gives them solid comfort...freedom
    MAILBOX.  JOHN'S
    DISPLAYING CHECK TO             ....peace of mind.
    ETHEL.  BOTH ARE
    SMILING THEIR
    SATISFACTION


86  PHOTO: EXT.                     Thanks to the security it provides,
    REPEAT SHOT 50
                                    there's been no need for them to

                                    give up their home.


87  PHOTO: EXT.                     It's theirs to enjoy, now more
    MS, HORTON FRONT YARD,
    HOUSE IN BACKGROUND.            than ever...a gathering-place for
    JOHN (ARM ABOUT ETHEL)
    IS HAPPILY WAVING               friends and family.
    GOODBYE TO BOB, DOROTHY
    AND THE GRANDKIDS


88  PHOTO: INT.                     Since they didn't have ~~them~~ to move
    REPEAT SHOT 58
                                    in with Bob and Dorothy, the Hortons

                                    create no friction in their son's

                                    home.


89  PHOTO: EXT.                     Instead, financially secure,
    MS, ENTRANCE TO SON'S
    HOME.  BOB AND DOROTHY          they're welcome guests--urged to
    AND THE GRANDKIDS ARE
    WELCOMING JOHN AND              drop by more often!
    ETHEL ENTHUSIASTICALLY

90　PHOTO: EXT.
　　REPEAT SHOT 46

With money coming in regularly
there wasn't any need for them to
sell their car.

91　PHOTO: EXT
　　MS, JOHN AND A COUPLE
　　OF FRIENDS IN FISHING
　　GARB IN SAME BOAT AS
　　IN SHOT 30.　THIS TIME,
　　HOWEVER, THE BOAT HAS
　　A SHINY NEW MOTOR AND
　　JOHN IS PROUDLY GUNNING
　　THE CRAFT AWAY FROM THE
　　DOCK

--When fishing season opens, John
even can afford a motor for his
boat!

92　PHOTO: EXT.
　　REPEAT SHOT 32

Those scenic trips--?

93　PHOTO: EXT.
　　MCU, JOHN AND ETHEL
　　BOARDING PLANE

You bet they'll go--because
travel's in their budget.

94　PHOTO: EXT.
　　MS, MIDDLE-AGED MAN
　　INTRODUCING YOUNGER
　　MAN TO JOHN AND ETHEL
　　ON DOWNTOWN STREET.
　　JOHN IS SHAKING HANDS

Each day brings new people to
meet...

95　PHOTO: EXT.
　　MS, JOHN AND ETHEL
　　LAUGHING AND TALKING
　　WITH COUPLE OF SIMILAR
　　AGE IN BACK YARD OF
　　NON-HORTON HOME.　MAN
　　IS BROILING HAMBURGERS
　　OVER OUTDOOR FIREPLACE

...the warm glow of continuing
friendships...

Candles--20

96  PHOTO: INT.                    ...community respect on the level
    MS, CONFERENCE ROOM IN
    SHOT 16.   MEN ARE            reserved for the stable and the
    STANDING BESIDE CHAIRS,
    JOHN THIS TIME AT HEAD        solvent.
    OF TABLE.   SOME OF THE
    MEN ARE THE SAME AS
    BEFORE, BUT OTHERS HAVE
    BEEN REPLACED BY NEW
    FACES

97  PHOTO: EXT.                    The very same years that have
    MCU, POORLY-DRESSED
    OLD MAN, DEPRESSED            brought so many men and women
    EXPRESSION, SITTING
    SLUMPED ON PARK BENCH         down to the "old and out of a

                                   job" category...

98  PHOTO: EXT.                    ...have seen John and Ethel firmly
    CLOSE MS, PARK ACTIVITY
    AREA.   JOHN AND ETHEL        established as happy, honored
    ARE LAUGHING AND
    PLAYING SOME GAME IN          senior citizens.
    FOREGROUND, WITH GROUP
    OF FRIENDS

99  PHOTO: INT.                    Well, thanks to John's foresight,
    REPEAT SHOT 6
                                   that's the way it worked out for

                                   the Hortons.

100 ART: ILLUSTRATION              But the self-same need for income
    SILHOUETTED FAMILY
    GROUP--MAN, WOMAN,            arises in every home--simply
    TEEN-AGE BOY AND GIRL,
    BOY OF EIGHT OR TEN           because sooner or later earned

                                   income stops for every man--but

                                   the need for money goes on the

                                   same.

Candles--21

101 ART:(CONT.)
    CU, MAN'S HEAD IN
    PROFILE, SILHOUETTED
    AS HE PONDERS WITH
    hAND ON CHIN

And when that day arrives, as it must, what happens to the bread-winner and those dependent on his earning power?

102 ART: (CONT.)
    SOMBER, DEPRESSED MOOD
    MAN IN SILHOUETTE,
    CRINGING FEARFULLY
    AND DOMINATED BY
    LOOMING BACKGROUND

Will the result be loss of income --humiliation--bitterness--regret --charity?

103 ART: (CONT.)
    BRIGHT, OPTIMISTIC
    MOOD MAN IN SILHOUETTE
    TOWERING DOMINANT OVER
    SUBORDINATED BACKGROUND

Or, will intelligence and forethought intervene, to provide a guaranteed, livable income through all the golden years...

104 ART: (CONT.)
    MAN IN SILHOUETTE IN
    LOUNGE CHAIR, FEET ON
    FOOTSTOOL,  RELAXED
    AND EXPANSIVE

...a chance for the man of the house to enjoy to the full the happiness he's earned?

105 ART: (CONT.)
    SOMBER, DEPRESSED MOOD
    MAN'S SLUMPED FIGURE IN
    SILHOUETTE, ISOLATED
    AND FORLORN

Will he "just scrape by", eking out a meager, worry-racked existence on crumbs from the table of life...

106 ART: (CONT.)
    REPEAT SHOT 4

...or will there, as long as he lives, be candles on his birthday cake?

Candles--22

107 ART: (CONT.)          This is the decision every man
    REPEAT SHOT 101
                          must make for himself and for

                          his family.


108 ART: (CONT.)          Which will it be for you and
    REPEAT SHOT 100
    SUPERIMPOSE TITLE:     yours?
    "WHICH WILL IT BE FOR
    YOU AND YOURS?"


                          (MUSIC: UP TO FINALE)

single-screen slide show, except for more painstaking planning and calculation. But sometimes problems arise, and you need to think twice about what's involved before you commit yourself. Keep in mind Bill Pryor's comment in Chapter 5 about keeping it simple.

## FILMSTRIPS

Closely related to the slide show is the *filmstrip,* a length of 35mm film on which a series of still photos or graphics are printed in sequential order. Ordinarily including from 20 to 50 frames, it may be in either black-and-white or color, silent or accompanied by sound. In effect an illustrated lecture, the filmstrip permits a pause of any length for any frame, plus the opportunity to change or expand the narration to fit special needs. Except for the packaging of the pictures and the means of projection, there's very little difference between a filmstrip and a slide show.

In the eyes of the users, however, these differences can be great. The slide show costs more (50 slides run markedly higher than an equivalent length of film), weighs more, occupies more space, and is more difficult to handle and ship. The filmstrip, in contrast, weighs next to nothing, can be stored in a can an inch in diameter and an inch-and-a-half high, and can be shipped neatly in any film mailing envelope. These are all understandable deciding factors if you are buying 10,000 copies of a given presentation.

The projection equipment, by and large, can be called separate-but-equal, with personal preference playing a large role. As for sound, both slides and filmstrips in most cases use tape cassettes or microgroove or electrical transcription discs.

Where the writer is concerned, however, all this tends to be pretty much irrelevant. The issue always comes back to the same basic principles we have talked about before. They apply, whether the writer's scripting a slide show or a filmstrip.

To get down to specifics, here are some comments by Rachel Stevenson, Director of Filmstrips, International Film Bureau, Inc. In her *A Check List for Writers and Producers of Educational Filmstrips,*\* she says:

1.  How do you choose a subject for an educational filmstrip? Look over a school system's curriculum of study. Choose a subject that will fit into the curriculum. Test your chosen subject by discussing it with educators to see whether it would be helpful.
2.  Determine the purpose of your filmstrip and keep this purpose well in mind during production.
3.  Before writing the script, work out a theme which will be developed. Outline *how* this theme will be developed.

---

\*Excerpt from *A Check List for Writers and Producers of Educational Filmstrips* used by permission of Rachel Stevenson and International Film Bureau, Inc.

4. Plan how the visuals can develop the theme.
5. Of course, be sure all facts are accurate.
6. Write with a direct, clear style. Keep in mind the vocabulary of the level of the audience.
7. Avoid just a list of pictures. Have the script with its visuals develop the theme and reach a conclusion.
8. A summary may often be helpful.
9. Use your visuals; let them tell the story. Let the picture speak for itself.
10. The first sentence in a new frame should tie in with or identify with the picture. For instance, if the conspicuous part of the picture is a horse in the foreground, don't talk about unrelated information and mention the horse at the end of the frame; people will wonder why they are looking at a horse.
11. It is very important to have only one idea in a frame with a clear visual to illustrate it.
12. The visual should read from left to right. If necessary change the script so that both the script and visual read from left to right. If there is no printing in the picture, it is sometimes possible to turn the picture over to make it read from left to right with the script.

Although Stevenson's remaining points apply primarily to the work of the producer, they are also the kind of thing you may find worth remembering—either because you some day may find yourself in the role of producer or because these are issues sufficiently important to the success of a filmstrip that you may want to raise your voice a bit rather than see a job go sour through no fault of your own.

13. Often sound and appropriate music enhance a filmstrip. But never let the sound or music detract from the narration.
14. It is often a delicate matter to keep the sound and music at just the right level. It must never drown out the narration, but if too low it will be lost when played on most school equipment that is not hi-fidelity.
15. Take pains when bringing in the audible bell for manually advancing the filmstrip. Sound appropriate for one frame but not for the following frame should end as the bell (or automatic advance) brings in the next frame. Failure to do this accurately results in a sloppily produced filmstrip.
16. Take pains in recording the bell signal. Keep the volume low enough so as not to be startling, but be sure it can be plainly heard. Nothing exasperates a teacher or student more than to get the filmstrip behind the audio. Be sure to lower the music if necessary between frames to make the bell audible. Avoid recording the bell on a note of music that is identical with the tone of the bell. It is surprising how often this situation arises.

These are all solid points—practical professional advice with years of experience in the field behind it.

Look at the excerpts from an International Film Bureau script, *Grant Wood*, to see how all this shapes up in script form.

Rachel Stevenson, the scriptwriter, director, and producer for *Grant Wood*, provides insight into the creative process:

music: major 6066 B-6 *Opening*

Start music
24 seconds from start        (Black frame)
                             (piano is out + flutes are in)

at 26 seconds 1st title is —

International Film Bureau Inc.
presents

GRANT WOOD
HIS LIFE AND PAINTINGS

Copyright c MCMLXXXIV International Film Bureau Inc.
All Rights Reserved                                    13

at 36 seconds from start 2nd title.

GRANT WOOD'S LIFE
                                                        7
Part 1
                                                        2nd.
make a very slow fade.

on first picture no music
                                                        27

                                                        7

                                                        17

                                                        24 FI

(Courtesy of Wesley H. Greene, President, International Film Bureau, Inc.)

With appreciation

to the many friends of Grant Wood
who gave us their time and information,

to the many private collectors who have
made our series possible,

And to the following museums:

The Art Institute of Chicago
The Minneapolis Institute of Arts
Davenport Art Gallery
The Cincinnati Art Museum
The Metropolitan Museum of Art, New York
Amon Carter Museum, Ft. Worth
Hirschl and Adler Galleries, Inc., New York,
    and James Maroney, New York
The Des Moines Art Center
Cedar Rapids Museum of Art

writer-producer-director
RACHEL STEVENSON

photographer
EGONS TOMSONS

editor
BEN HODGE

Distributed by

INTERNATIONAL FILM BUREAU INC.

Chicago, Illinois

*6*

International Film Bureau Inc.
presents

GRANT WOOD
HIS LIFE AND PAINTINGS

Copyright c MCMLXXXIV International Film Bureau Inc.
All Rights Reserved

*16.*

GRANT WOOD'S PAINTINGS

Part 1

*25*

I

4 end titles

13.47

*Engineer*
*rw.*

| GRANT WOOD'S LIFE PART 1 | GRANT WOOD<br>HIS LIFE AND PAINTINGS |
|---|---|
| 1. American Gothic | "American Gothic" is the painting most readily associated with the artist, Grant Wood.  When it was exhibited in the Art Institute of Chicago in 1930 its reception was sensational. |
| 2. Sketch | This is Wood's first sketch for the painting.  Note that at this time the man held a rake instead of a pitchfork. |
| 3. Painting of Gothic House | This is Wood's painting of the small white house with its Gothic window, which inspired "American Gothic." |
| 4. Full Am. Goth. | Wood described his idea for the painting: "I imagined Gothic people with t[ ] faces stretched out long to go with the house.  Any northern town old enoug[ ] to have some buildings dating back to the Civil War, is liable to have a house or church in the American Gothic style.  I simply invented some American Gothic people to stand in front of a house of this type." |
| 5. CU two heads | Wood's sister, Nan, was good enough to pull her hair back primly and pose as the austere daughter.  For the equally austere father, Wood looked around for a model. |
| 6. Full shot | One day in his dentist's chair he stared at Dr. McKeebe, then finally said, "I like your face."  McKeebe, a little startled, went on with his work. |
| 7. CU hand with pitchfork | Later, Wood grabbed McKeebe's hand, turned it over and remarked, "Now there's a hand that can do things."  The hand Wood admired is the one that holds the pitchfork. |
| 8. Photo | This photograph of Grant's sister, Nan, and Dr. McKeebe was taken twelve years after "American Gothic" was painted.  Grant had assured their anonymity, but by this time they were known as the models. |

life 1                                    2

9. "Young Corn"     Next to "American Gothic", Grant Wood is probably remembered most for his

Iowa landscapes, such as "Young Corn", with their rolling hills,

plowed fields, and impressionistic trees.          Wood "found himself"

when he began to paint things he knew in Iowa.

o

10. CU feathers     Even as a child he liked to paint.  One day, early in his life, he was

painting rows of crescents and when his mother asked about it, he said he was

painting a chicken.  The crescents, of course, were rows of feathers.  He

might have painted this under the dining room table hidden by a long checked

[Grant's childhood     tablecloth where he often found privacy from his sister and two brothers.
painting]
11. Grant at 10     When Grant was only ten, his father died, and Mrs. Wood had to sell the

farm and take her children to live in Cedar Rapids near relatives.

12. Daffodils       In school, Grant often painted what the class was studying.  Here he

illustrated a passage from Wordsworth's "Daffodils."

13. Boy in Raincoat Wood kept some of his childhood paintings.  Later in life, he told

parents, "It is as natural for children to draw as it is for them to

breathe.  They are very serious about their earliest efforts.

14. Cat and Rug     "Don't criticize these early gropings.... They are important to the children

who make them.  I was as bashful a child as ever lived... But I would pour

out all my emotions and longings in a painting and my mother understood and

encouraged me.

15. Beets           "Everyone should experience this joy of creation."  This advice might

have been prompted by an incident supposed to have happened during Grant's

boyhood.

36. Wood in     In 1918 Wood was in the army.  Finding time now and then to sketch, he
    Uniform
                did portraits of enlisted men for 25 cents but charged officers a dollar.

37. "Old Sexton's   "Old Sexton's Place" is one of the paintings Wood did when he returned
    Place"
                    to Iowa in 1919.  At this time he had no source of steady income.

38. Miss Prescott   Miss Prescott, principal of Jackson Junior High School, gave Wood a
    1919
                    part-time job teaching art, even though she thought he was too shy to

                    make a good teacher.  Here's Miss Prescott.

                    (TAPED VOICE OF PRESCOTT)

*[I was lucky to find this tape at a school where Mrs. Prescott used to teach]*

*(reprise ...)*

                    "Oh my, I was so mistaken--he was just perfectly

                    marvelous with the children, and the children were just

                    beside themselves about him, from the beginning he was

                    just like the Pied Piper, when they saw him coming,

                    there'rd be just an army of them run down 4th Avenue to

                    meet him and they'd pelter him with questions.  He always

                    had not only had an intelligent--he was well read--an

                    intelligent answer but he'd have an entrancing sense of

                    very quiet humor.  He'd make them laugh and tell them

                    funny things."

41. McKinley
    School

In 1922, Wood transferred along with Miss Prescott to McKinley Junior

High School.

42. Mrs. Kesler
    outside room

Mrs. Carl Kesler recalls those days.

(TAPED VOICE OF KESLER)

[It helps to have contemporary people recall the past. I did this several times in the script.]

"Yes, this was Grant Wood's classroom.  I was in the

8th grade when Grant was teaching here.  I didn't take

art, but I remember how my friends liked him, and I

would look in his room and see all the fun things on the

wall, and sometimes when I'd pass this room I heard

a lot of noise going on in there."

43. Boys in 9th
    grade.

The boys in Wood's 9th grade class were making the noise.  When

complaints reached Miss Prescott's office, Wood found a solution.  Miss

Prescott tells about it.

44. Mourner's
    bench

(TAPED VOICE OF PRESCOTT)

"So he designed this little..what we called the
Mourner's Bench.  He designed it and the children

went to the manual training shop and built it.

45. Crying head

And he carved the..those heads and the lettering,

he himself did that."

replace

47. Full bench

The inscription reads:  "The Way of the Transgressor is Hard."--The bench

was in Miss Prescott's outer office where pupils who misbehaved had to sit.

Wood's students were proud of making it and made sure they never had

to sit on the bench.

c

| | |
|---|---|
| 48. Wide shot class | Another project was painting a 150-foot frieze on paper for the school cafeteria.  The frieze told the story of a trip to the Imagination Isles.  Wood wrote:  "Our hosts on this trip are forty-five ninth grade boys--they have used only the highest grade of oil paint and the very best imagination that is possible to be found locally." |
| 49. CU boy at work | Each student had 3 or 4 feet to decorate. |
| 50. Full shot | About the frieze's theme Wood wrote, "Almost all of us have some dream-power in our childhood, but without encouragement it leaves us and we become bored and tired and ordinary.  Then someday when we are physically comfortable we remember dimly a distant land we used to visit in our youth.  We try to go again but we cannot find the way." |

At first, we thought of making a film on Grant Wood. I went to Cedar Rapids, Iowa, and other [of] Grant Wood's country areas and made notes on shooting sites and angles. But I soon realized that Wood was such an interesting person and his art so fascinating that I couldn't do him or his paintings justice in a thirty-minute film. I could get much more material in a series of filmstrips.

To that end, I read everything I could find about Grant Wood. I talked to people in Cedar Rapids who had known him well.

Finally, I decided to divide my script into four parts. Two would include paintings that had a special influence on his career—even paintings he made as a child. Two more scripts would cover more of his paintings, with the history and story behind each one. More analysis would appear in these scripts. I never forgot that the script is the most important part of a production.

Many people conducting such a project would have taken a wide shot of each picture; then they would have made their close-ups of details in their workshops from their wide shot. We did not do this. I was so anxious to have the color and sharpness as perfect as possible for this art filmstrip series that for every close-up (and there were many), we reset the camera, changed the lighting and lens, and shot the detail firsthand from the original painting.

In a series like this one, and in many documentaries, interviews with people associated with the subject add authenticity and interest. In Cedar Rapids I recorded a statement about the first painting Grant Wood sold. It was when he was a boy. He sold it for eight dollars—which was the sum he needed to buy supplies to paint more pictures. The series contains several such recorded statements.

In the following pages are two more filmstrip script formats, courtesy of the U.S. Postal Service Training and Development Institute. The first, a common type, includes a storyboard frame for each shot on the script's visual side, though it is used more for notes than art in our sample.

The second example, simpler and less formal, replaces the storyboard with words, telling what art is to be inserted to go with the narration.

## MULTI-IMAGE DISPLAYS

Traditionally, *multi-image* referred to a number of different slides being projected simultaneously side by side on a screen. Multiple projectors were used and were coordinated to work together. Sound was synchronized to accompany specific images.

In today's multi-image display, adjacent slides will sometimes blend together to form a larger image, and sometimes every slide will depict a different scene. Sometimes two or more projectors are focused on a single screen; sometimes many projectors are used, with each projector focused on a different screen. It's a matter to be decided.

Today, too, while *multi-image* refers to many still images projected simultaneously, the technology has changed. Though a bank of projectors

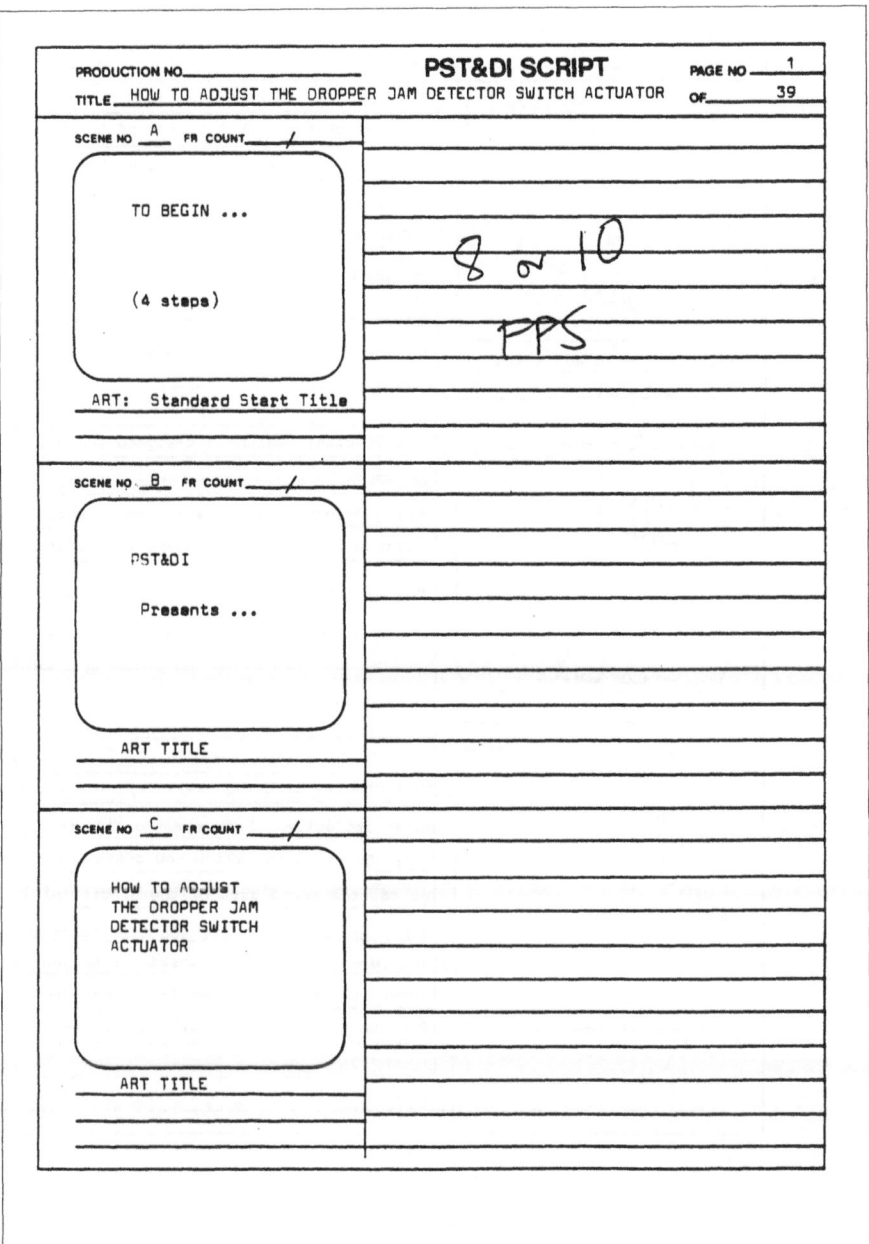

Excerpt from filmstrip script. (Courtesy of the U.S. Postal Service Training and Development Institute.)

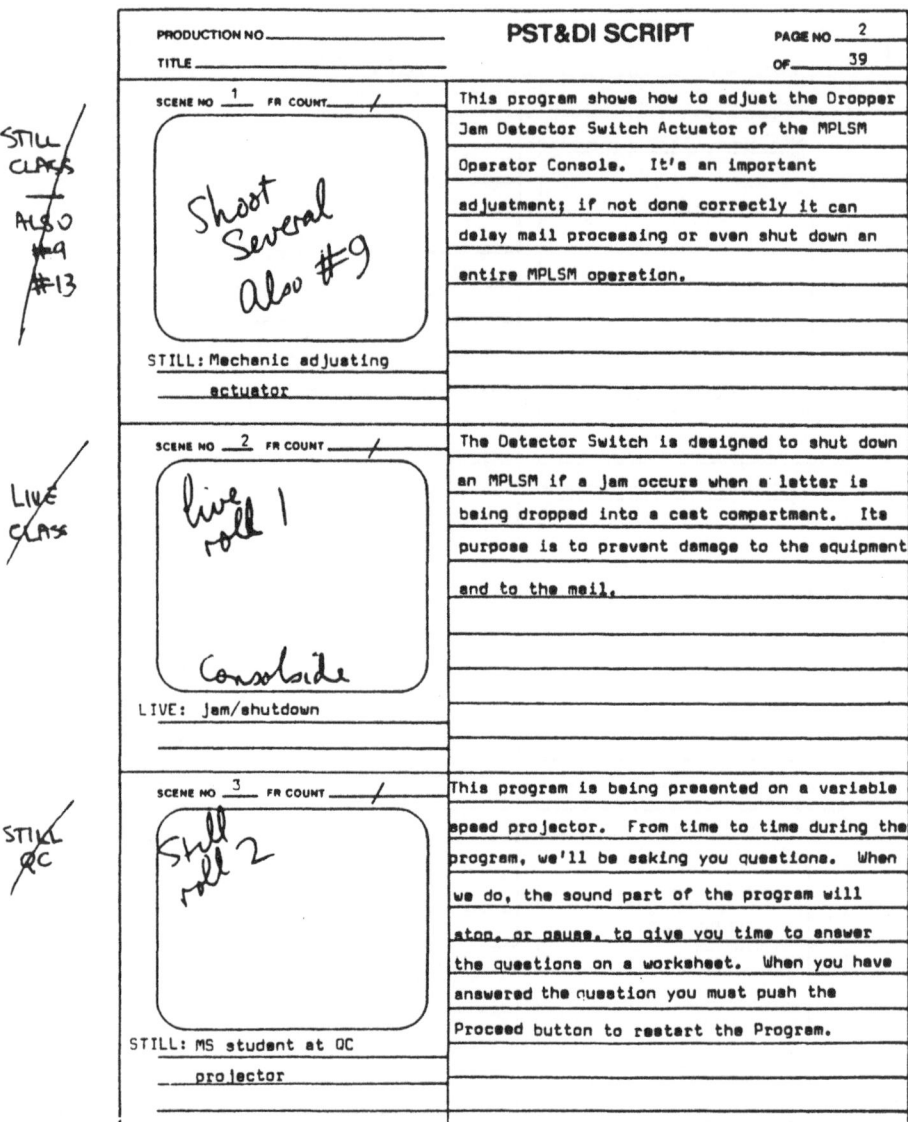

STILL/
CLASS

ALSO
#9
#13

LIVE/
CLASS

STILL/
QC

PST&DI SCRIPT

PRODUCTION NO _____
TITLE _____

PAGE NO ___2___
OF ___39___

SCENE NO __1__   FR COUNT ___/___

*Shoot Several also #9*

STILL: Mechanic adjusting
        actuator

This program shows how to adjust the Dropper
Jam Detector Switch Actuator of the MPLSM
Operator Console.  It's an important
adjustment; if not done correctly it can
delay mail processing or even shut down an
entire MPLSM operation.

SCENE NO __2__   FR COUNT ___/___

*live roll 1*

*Consolside*

LIVE: Jam/shutdown

The Detector Switch is designed to shut down
an MPLSM if a jam occurs when a letter is
being dropped into a cast compartment.  Its
purpose is to prevent damage to the equipment
and to the mail.

SCENE NO __3__   FR COUNT ___/___

*Still roll 2*

STILL: MS student at QC
        projector

This program is being presented on a variable
speed projector.  From time to time during the
program, we'll be asking you questions.  When
we do, the sound part of the program will
stop, or pause, to give you time to answer
the questions on a worksheet.  When you have
answered the question you must push the
Proceed button to restart the Program.

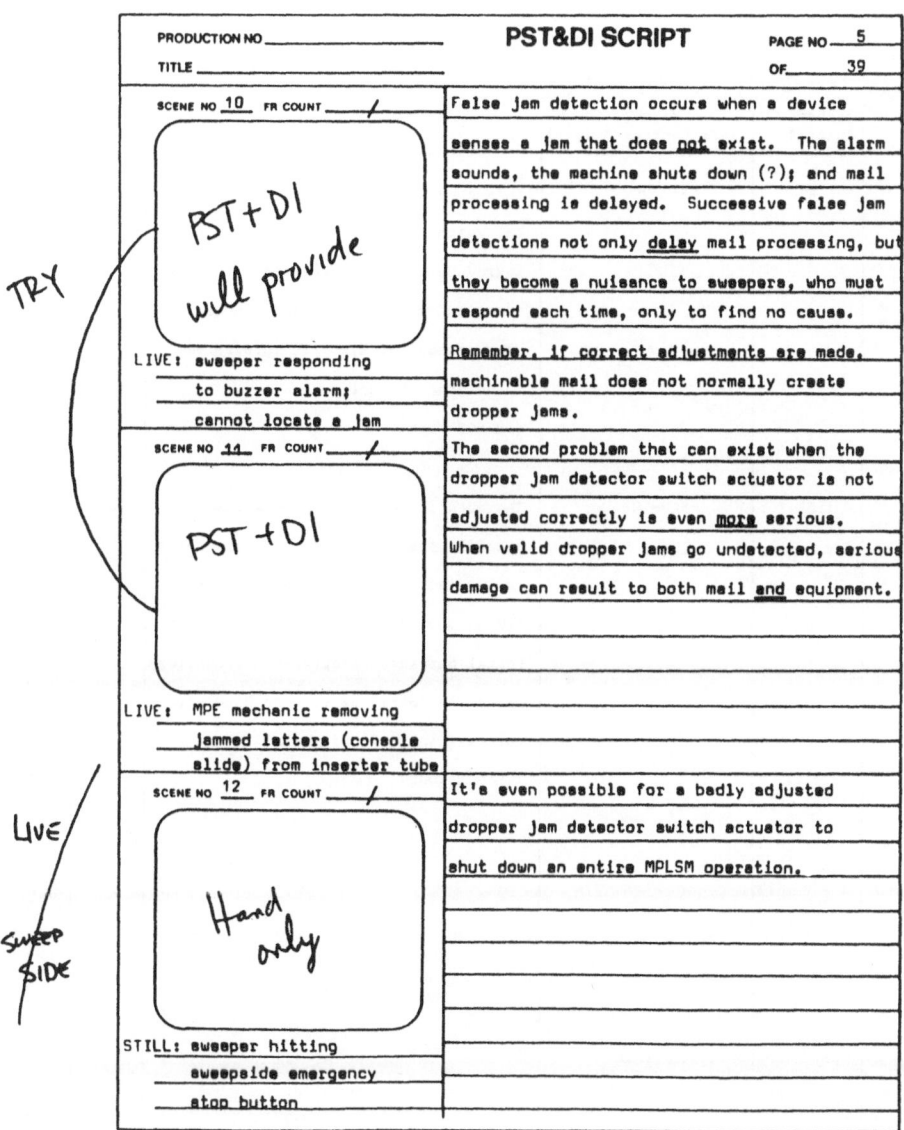

PRODUCTION NO _____

TITLE _____

PAGE NO ___5___

OF ___39___

**SCENE NO 10** FR COUNT ___/___

*PST+DI will provide*

*TRY*

LIVE: sweeper responding
      to buzzer alarm;
      cannot locate a jam

False jam detection occurs when a device
senses a jam that does not exist.  The alarm
sounds, the machine shuts down (?); and mail
processing is delayed.  Successive false jam
detections not only delay mail processing, but
they become a nuisance to sweepers, who must
respond each time, only to find no cause.

Remember, if correct adjustments are made,
machinable mail does not normally create
dropper jams.

**SCENE NO 11** FR COUNT ___/___

*PST + DI*

LIVE: MPE mechanic removing
      jammed letters (console
      slide) from inserter tube

The second problem that can exist when the
dropper jam detector switch actuator is not
adjusted correctly is even more serious.
When valid dropper jams go undetected, serious
damage can result to both mail and equipment.

**SCENE NO 12** FR COUNT ___/___

*Hand only*

*LIVE*

*SWEEP SIDE*

STILL: sweeper hitting
       sweepside emergency
       stop button

It's even possible for a badly adjusted
dropper jam detector switch actuator to
shut down an entire MPLSM operation.

SIMM Audio Visual Script

| Sequence Identifier | Visual (Art Control Number) | Narrative | Time (Sec) |
|---|---|---|---|
| 1A1 | SAV-200 | This program has been developed for the United States Postal Service to introduce you to a new kind of technical manual, the SIMM manual... SIMM stands for Symbolic Integrated Maintenance Manual. (CHANGE TONE) | 15 |
| 1B1 | SAV-201 | The quantity of mail has increased steadily every year in the recent past. Mail volume will continue to increase, and so must the efficiency of the U.S. Postal Service. Progress in technology helps the Postal Service to handle this constant increase of mail. (CHANGE TONE) | 10 |

Excerpt from the Simm Audiovisual script. (Courtesy of the U.S. Postal Service Training and Development Institute.)

SIMM Audio Visual Script

| Sequence Identifier | Visual (Art Control Number) | Narrative | Time (Sec) |
|---|---|---|---|
| 1C1 | SAV-202 | But progress in technology means we have to learn to use increasingly complex machines and devices to perform faster and more efficiently the operations which used to be done by hand. (CHANGE TONE) | 13 |
| 1D1 | SAV-202 ① | Machines and devices can't work without skilled men and women. They need us to operate and maintain them... Our success depends upon people using the machines efficiently. (CHANGE TONE) | 12 |

still may be used, now the images often are recorded on a single videotape, on a disc, or in a computer, and projected from one source.

Because the field is complicated and confusion is not at all uncommon, it's highly desirable that you begin by finding out what the person to whom you're talking means.

Script preparation in multi-image tends to follow the pattern already established: What are the client's needs? What do I want to say? Who's the audience?

Beyond this, the approach, length of presentation, and costs need to be determined. Perhaps research needs to be done, a flow chart laid out.

Finally, a treatment or script is written, as outlined in previous chapters. A story line provides the program's connective thread and ties the ideas or theme together in some logical order. Although other ideas or cross threads may be interwoven in the fabric, the story line carries the dominant thought or theme throughout.

A storyboard may or may not be prepared. Such a series of sketches or images visually represent what the producer intends to portray. It shows the composition of each frame in a sequence, as well as the continuity between frames and sequences.

For the script format, whatever comes handy is the rule. The script doesn't even have to specify the video shots to be included. Laying out the audio is considered sufficient.

An involved example is this bit from a PDC Media Productions script. Again, narration is written first, but it's laid in against a script format sheet that provides space for shots from three or more projectors, if desired.

In other words, the presentation is infinitely adaptable. The challenge is merely deciding what shall be used.

## VIDEO WALLS

Closely related to both the slide show and linear video is the video wall. An array of video monitors usually is arranged so that the overall display retains the TV image format. In its simplest form it's a big picture display, using an electronic "splitter" to divide a single video signal across the required number of monitors.

Video walls provide an element of spectacle—what one writer calls "visual razzle-dazzle"—to retail shopping or any situation that brings people together. It's particularly popular for conventions and the like. It may add interest to a reception room or lobby. Its visual effects are stunning, with a picture brightness not possible with projected images.

"We use it as a vehicle to entertain our customers and to provide them with product information about the things we sell in the store," says Sue Sorenson, public relations director of Dayton's Department Store, Minneapolis. "The programming that's produced for the wall varies—everything from live fashion shows that we put on ourselves to actual TV

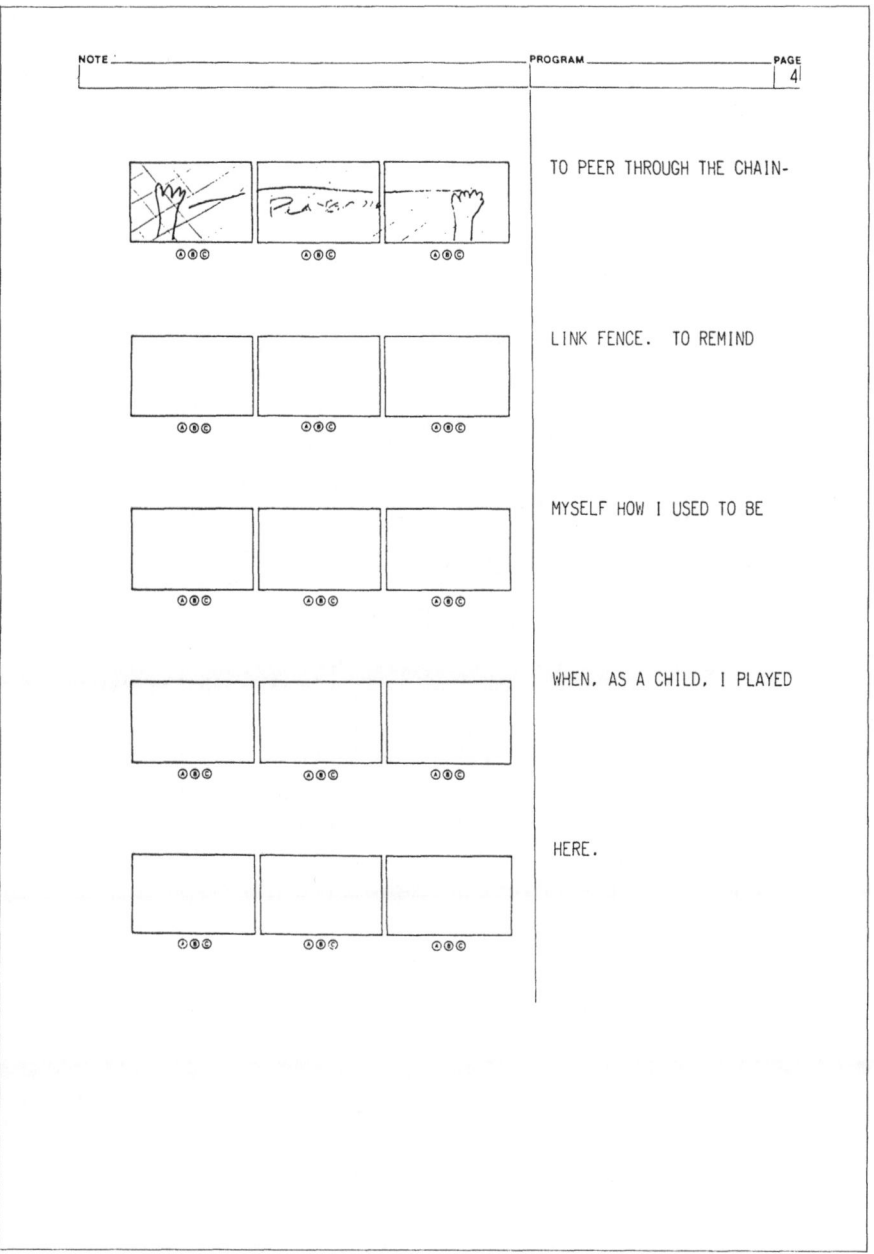

(Courtesy of PDC Media Productions.)

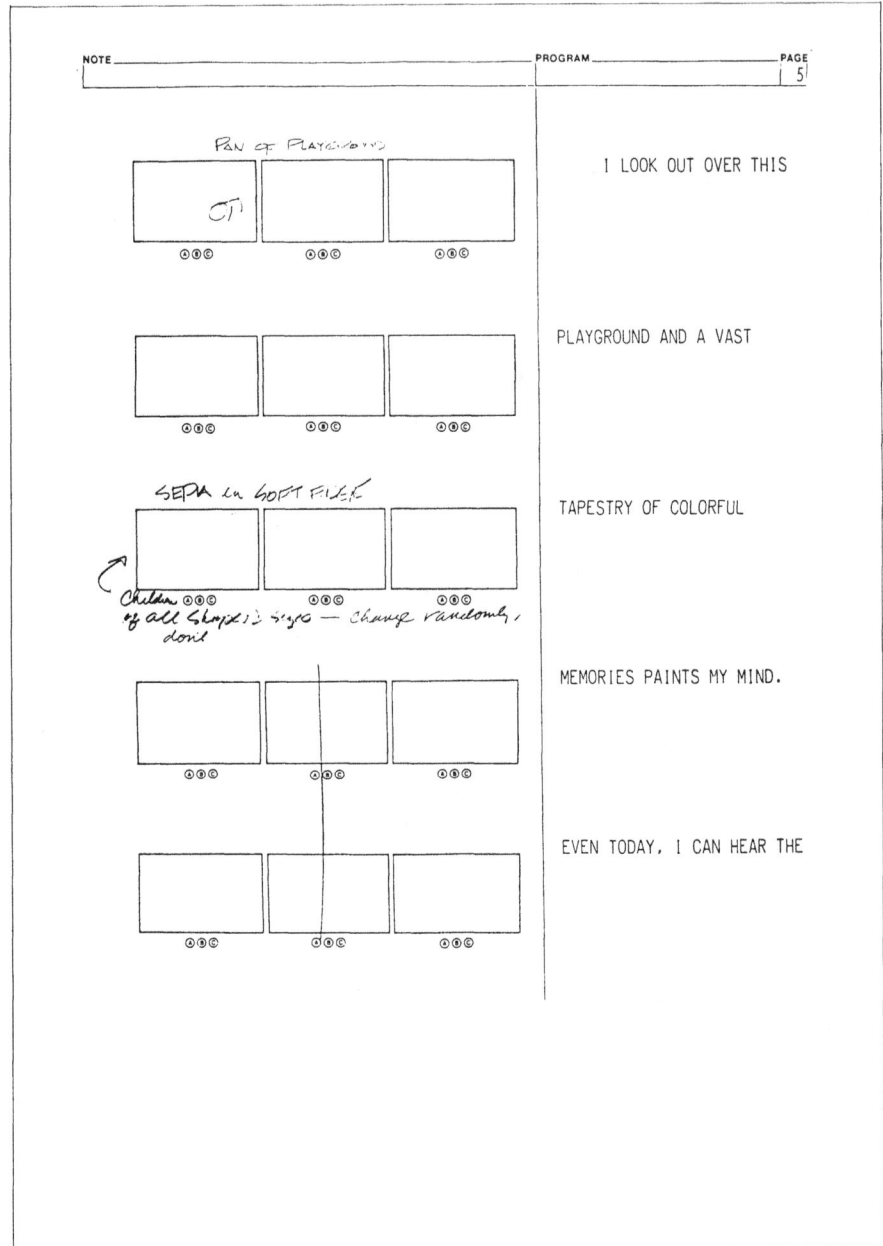

NOTE _____  PROGRAM _____ PAGE 5

PAN OF PLAYGROUND

CT

I LOOK OUT OVER THIS

PLAYGROUND AND A VAST

SEPA in SOFT FILER

Children of all Shapes & sizes — change randomly, don't

TAPESTRY OF COLORFUL

MEMORIES PAINTS MY MIND.

EVEN TODAY, I CAN HEAR THE

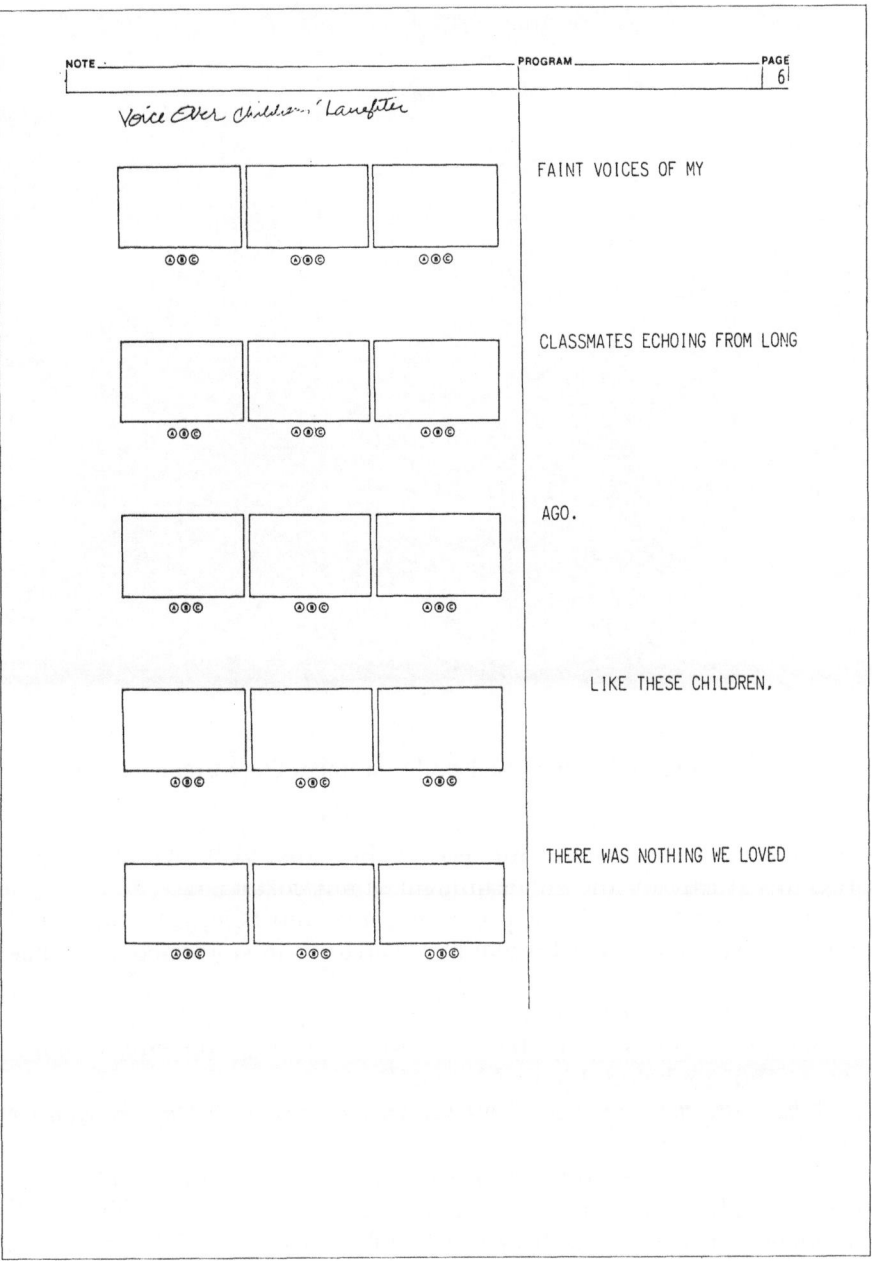

NOTE

*Voice Over Children's Laughter*

PROGRAM                    PAGE
                            6

FAINT VOICES OF MY

CLASSMATES ECHOING FROM LONG

AGO.

LIKE THESE CHILDREN.

THERE WAS NOTHING WE LOVED

(Courtesy of Dayton Hudson Department Store Company.)

commercials that we adapt for use on the video wall. We also produce programs strictly for the entertainment of our customers."

How does a video wall program come into being? To answer that question, let's talk to Kristin Staubitz, Dayton's in-store media manager.

*What's your background, Kristin?*

*Staubitz:* I have a design background. I was an art director for a multi-image company before I joined Dayton's.

*What were the biggest problems you encountered in making the transfer to working with video walls?*

*Staubitz:* Well, it's a different medium, but the concept is similar. You're working with moving images rather than still images. Personally, I had no experience on a video wall before I took this job and I learned, so it's certainly doable.

*How do you go about putting a show together?*

*Staubitz:* I plan it and use our in-house staff or hire freelancers to execute it. When it's done I look at it and have them make changes if necessary. The arrangement of the image on the wall is sort of a subjective thing, much like any other multi-image show. You could program one particular tape many different ways and still have an effective presentation from our point of view.

*Does somebody write a script for a video wall production?*

*Staubitz:* It usually isn't scripted per se unless we are using a voice-over. I give them a single screen linear video and sometimes I'll have specific requirements—like I'll say, "You have to show the product name a certain number of times." Things like that. We find it too cumbersome to do a storyboard for each individual show—although I used to do that when I worked for a multi-image company. Our programs change at least weekly and I have anywhere from eight to twenty shows in one cycle.

*How is it programmed?*

*Staubitz:* Programming the wall is somewhat complicated. It's done with a computer. Each show is programmed independently. It's a technical process whereby a video signal is digitized. The computer commands the signal via the video wall hardware to go to a certain monitor at a certain time. It's done on the spot. We don't usually prepare a plan or script ahead of time because it's preferable to be able to see the image as you're programming. The video wall can magnify an image; it's a grid of TV screens. You may plan to use a two-by-two, which is how we refer to the size of the image. But when you see the two-by-two you may realize that the person's face is being cut in half by one of the grids on the wall and then you may decide it looks better as a three-by-three or as a single screen. You determine the size of the area that will show as one image and how each transition will look. That's the whole concept behind video walls in general.

*Do you have to be a computer programmer to do all this?*

*Staubitz:* You don't have to be a computer programmer. You must be familiar with this computer program, but they're not writing the program each time; they're using a software package available from a manufacturer. We use a software package called "C-Through" designed by a company called Electrosonic. It's a software package they wrote designed to go with the hardware they designed. Several companies produce and manufacture video walls and write software for them.

*How long does it take to prepare a program from the very beginning?*

*Staubitz:* It depends on the content. But the physical programming takes approximately four hours per finished minute of show time. Oftentimes we'll get only a week's lead time to produce a show from beginning to end.

*Is it all original work?*

*Staubitz:* Sometimes. But often we work with an existing video that is provided by one of our vendors. And we use a lot of computer-generated graphics too.

And that's the story on video walls. While on the face of it no script is used, the wall is created for a purpose. To help it fulfill that purpose, someone has to decide what will be included in the presentation. This conceptualization is where the writer, by whatever name, comes in. He or she must identify the content that will best convey the desired message. Still photos, video, computer-based material, and digital controls are involved.

It's all another fascinating facet of AV!

# 12

# Linear Video/Film

One of your first assignments in AV, in all likelihood, will be a linear video or film script—video, most likely, since film isn't nearly as common as video these days. So it's highly desirable that you know just what linear video is.

Video is "a system of recording and transmitting information which is primarily visual, by translating moving or still images into electrical signals" (*Videodisc and Related Technologies: A Glossary of Terms*). In other words, it's a way to take pictures electronically.

Widely used in broadcast television, video signals can be transmitted over high-frequency carrier waves, sent through cable on a closed circuit, or recorded on tape or discs. Generically, audio and other signals as well as visual are included.

*Linear* video, in turn, is video that goes in a straight line, from the beginning of a presentation to its end, at a preestablished speed and in a preestablished order. The audience merely watches. For an example, put your favorite movie on your VCR.

Linear video is a major part of the audiovisual field. The range of its uses is well-nigh incredible. From teaching, selling, and demonstrating to medical recording, corporate communications, and MTV, it's probably the most effective medium for showing an audience a setting they couldn't see otherwise, such as a volcano's mechanism or deep in the ocean or inside the heart.

It's also one of the best ways to record or sum up an event or subject: a marathon, a spring tornado, mathematical discoveries of the past century. It's also excellent for presenting a simple process that's always done pretty much the same way, when it's not important that the audience remember very specific and detailed information. Use it, for example, to show how something is done when you don't expect your viewers to be able to do it after the showing, but only to understand the process, as in "This is the way a house is built."

159

Increasingly, too, video is finding a place in retail merchandising, with projectors located here and there throughout a store. "It serves as an aid and a guide," explains David Snyder, Marshall Field's vice president and home fashion director. "Visitors can stay and watch the tape and find out why the designers did what they did. It's better than having someone there trying to walk them through." Currently he produces four-minute tapes for Trend House, the promotional area in Marshall Field's Chicago flagship store.

"I see a *tremendous* future in promotional video for retailers," he says. "The vital factor is having creative, imaginative people conceptualizing each production. Without them, it's likely to turn out boring and stupid."

Other merchandisers agree. One speaks of "the atmosphere boost that visual onslaught of POP [point-of-purchase] video provides." Another says, "We use it the way other people use a coat of paint."

How does a video program come into being? The United States Postal Service, a major producer of video programs, has detailed the process in a chart.

The most important and time-consuming portions of this process, the Postal Service notes, are scripting, preproduction (that's planning and preparation), and postproduction (editing and otherwise winding up the operation after the picture's been shot). As the graph shows, scripting constitutes 20% of the time involved.

Putting a linear video script together is very much like writing a film script. Whether you're dealing with fact or fiction, you begin by preparing a proposal or synopsis, expand it into an outline or treatment, and, finally, write the script itself, using one of the formats described in Chapter 10.

What does a linear video script look like? Here are some examples. First, a few pages from a script written by one of this book's authors for a treatment center for neurologically-damaged and otherwise handicapped children. It began with preliminary meetings with McCarty Center staff people to determine the objectives and production problems of the project, then was set down in written form in a project treatment.

More conferences followed. This time the director was very much involved. The script that evolved used a typical two-column format, as shown in the excerpt.

The crew then shot the video according to the script. Ordinarily a narration polish would have followed, but time pressure prevented it on this occasion.

For another more formalized example, refer to the Postal Service Integrated Logistic Support System's introductory videotape script.

You can see how differently the Postal Service script is handled, with every detail laid down. That a printed form is used to ensure that requirements are met tells the story.

Linear video can do just about anything, and it's finding wide acceptance for all sorts of purposes. Learning to write for it is well worth any audiovisual scriptwriter's time.

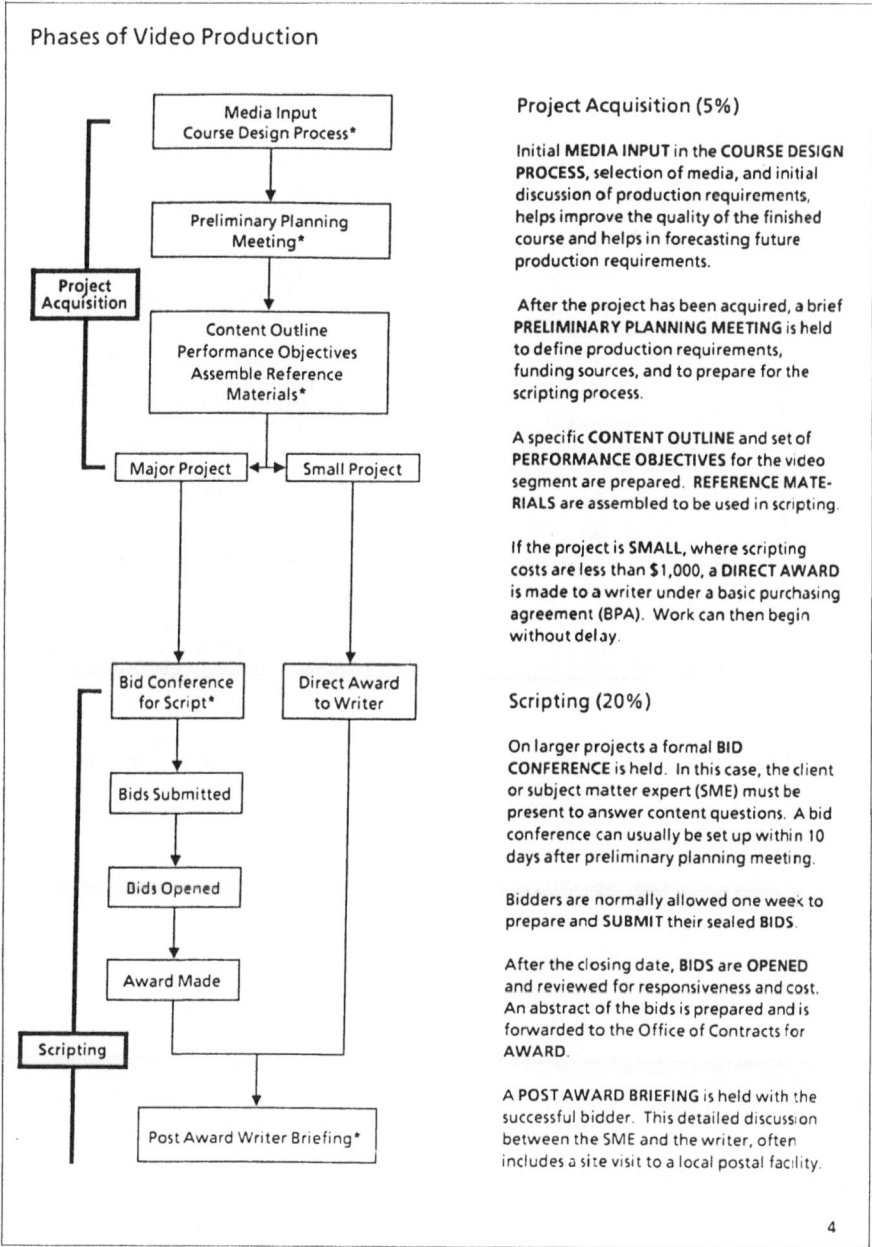

## Phases of Video Production

**Project Acquisition (5%)**

Initial **MEDIA INPUT** in the **COURSE DESIGN PROCESS**, selection of media, and initial discussion of production requirements, helps improve the quality of the finished course and helps in forecasting future production requirements.

After the project has been acquired, a brief **PRELIMINARY PLANNING MEETING** is held to define production requirements, funding sources, and to prepare for the scripting process.

A specific **CONTENT OUTLINE** and set of **PERFORMANCE OBJECTIVES** for the video segment are prepared. **REFERENCE MATE-RIALS** are assembled to be used in scripting.

If the project is **SMALL**, where scripting costs are less than $1,000, a **DIRECT AWARD** is made to a writer under a basic purchasing agreement (BPA). Work can then begin without delay.

**Scripting (20%)**

On larger projects a formal **BID CONFERENCE** is held. In this case, the client or subject matter expert (SME) must be present to answer content questions. A bid conference can usually be set up within 10 days after preliminary planning meeting.

Bidders are normally allowed one week to prepare and **SUBMIT** their sealed **BIDS**.

After the closing date, **BIDS** are **OPENED** and reviewed for responsiveness and cost. An abstract of the bids is prepared and is forwarded to the Office of Contracts for **AWARD**.

A **POST AWARD BRIEFING** is held with the successful bidder. This detailed discussion between the SME and the writer, often includes a site visit to a local postal facility.

4

(Courtesy of the U.S. Postal Service.)

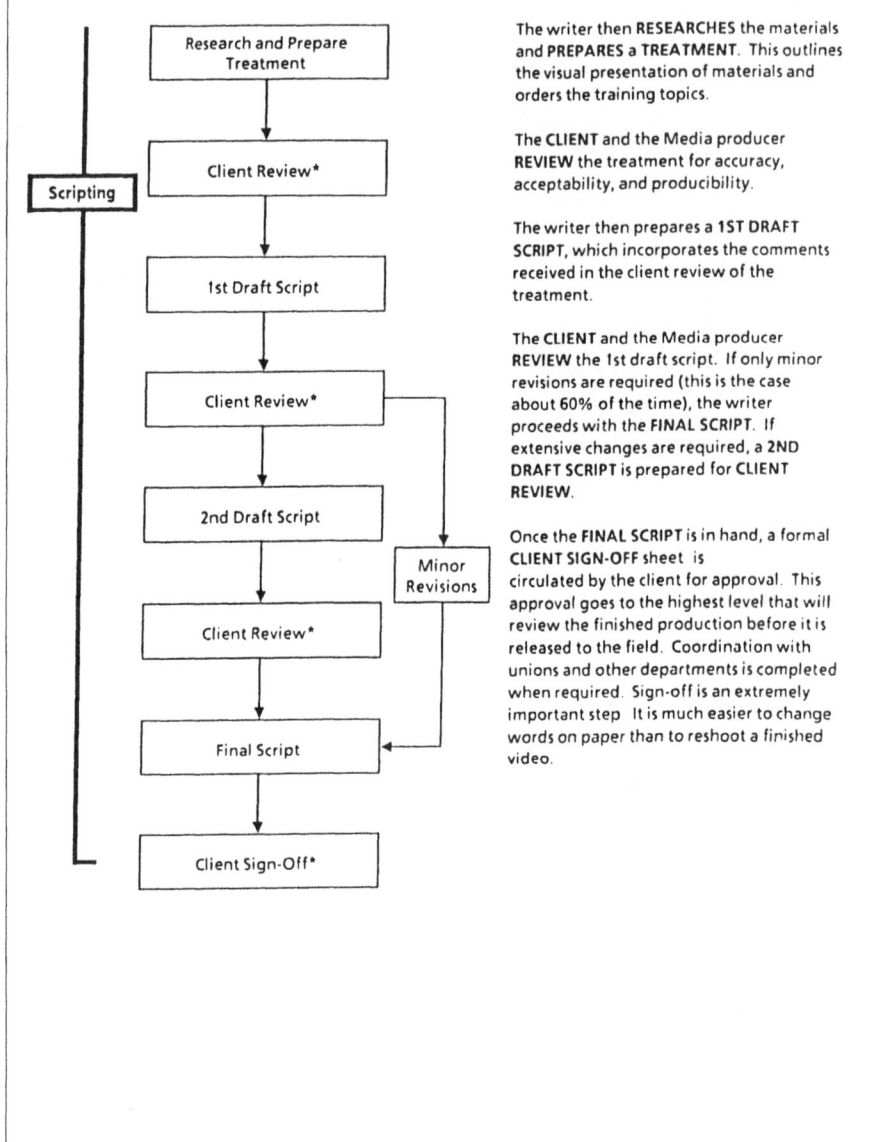

The writer then **RESEARCHES** the materials and **PREPARES** a **TREATMENT**. This outlines the visual presentation of materials and orders the training topics.

The **CLIENT** and the Media producer **REVIEW** the treatment for accuracy, acceptability, and producibility.

The writer then prepares a **1ST DRAFT SCRIPT**, which incorporates the comments received in the client review of the treatment.

The **CLIENT** and the Media producer **REVIEW** the 1st draft script. If only minor revisions are required (this is the case about 60% of the time), the writer proceeds with the **FINAL SCRIPT**. If extensive changes are required, a **2ND DRAFT SCRIPT** is prepared for **CLIENT REVIEW**.

Once the **FINAL SCRIPT** is in hand, a formal **CLIENT SIGN-OFF** sheet is circulated by the client for approval. This approval goes to the highest level that will review the finished production before it is released to the field. Coordination with unions and other departments is completed when required. Sign-off is an extremely important step  It is much easier to change words on paper than to reshoot a finished video.

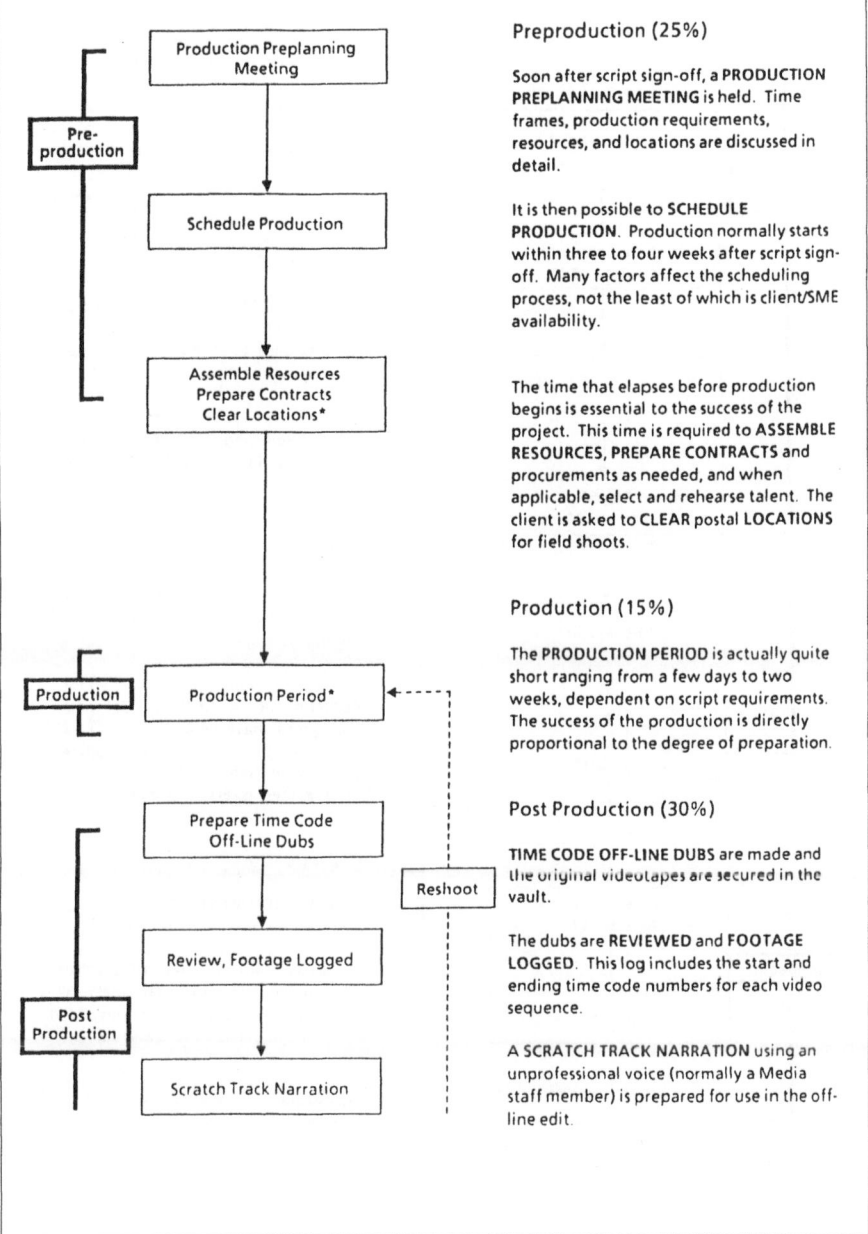

## Preproduction (25%)

Soon after script sign-off, a **PRODUCTION PREPLANNING MEETING** is held. Time frames, production requirements, resources, and locations are discussed in detail.

It is then possible to **SCHEDULE PRODUCTION**. Production normally starts within three to four weeks after script sign-off. Many factors affect the scheduling process, not the least of which is client/SME availability.

The time that elapses before production begins is essential to the success of the project. This time is required to **ASSEMBLE RESOURCES**, **PREPARE CONTRACTS** and procurements as needed, and when applicable, select and rehearse talent. The client is asked to **CLEAR** postal **LOCATIONS** for field shoots.

## Production (15%)

The **PRODUCTION PERIOD** is actually quite short ranging from a few days to two weeks, dependent on script requirements. The success of the production is directly proportional to the degree of preparation.

## Post Production (30%)

**TIME CODE OFF-LINE DUBS** are made and the original videotapes are secured in the vault.

The dubs are **REVIEWED** and **FOOTAGE LOGGED**. This log includes the start and ending time code numbers for each video sequence.

A **SCRATCH TRACK NARRATION** using an unprofessional voice (normally a Media staff member) is prepared for use in the off-line edit.

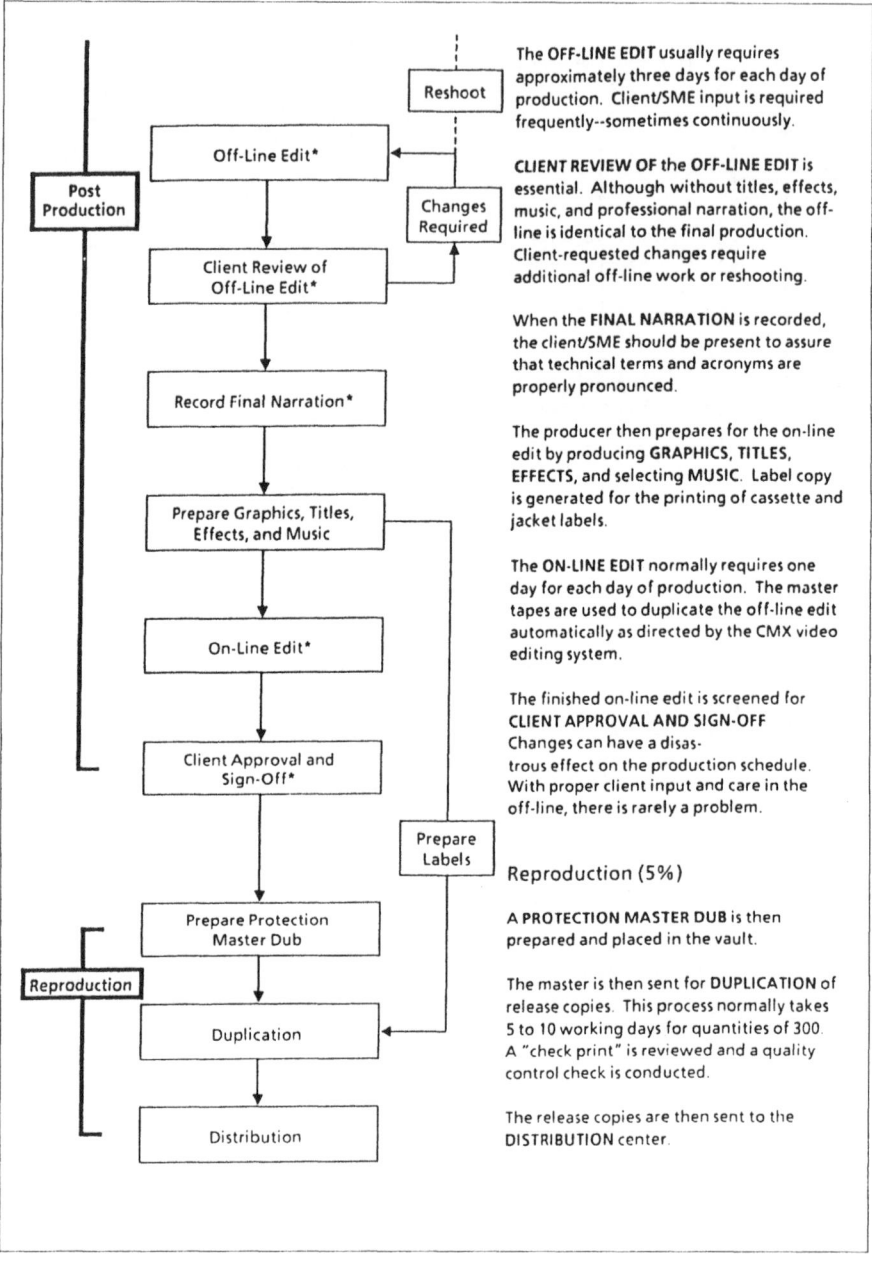

The **OFF-LINE EDIT** usually requires approximately three days for each day of production. Client/SME input is required frequently--sometimes continuously.

**CLIENT REVIEW OF** the **OFF-LINE EDIT** is essential. Although without titles, effects, music, and professional narration, the off-line is identical to the final production. Client-requested changes require additional off-line work or reshooting.

When the **FINAL NARRATION** is recorded, the client/SME should be present to assure that technical terms and acronyms are properly pronounced.

The producer then prepares for the on-line edit by producing **GRAPHICS, TITLES, EFFECTS,** and selecting **MUSIC.** Label copy is generated for the printing of cassette and jacket labels.

The **ON-LINE EDIT** normally requires one day for each day of production. The master tapes are used to duplicate the off-line edit automatically as directed by the CMX video editing system.

The finished on-line edit is screened for **CLIENT APPROVAL AND SIGN-OFF** Changes can have a disastrous effect on the production schedule. With proper client input and care in the off-line, there is rarely a problem.

Reproduction (5%)

A **PROTECTION MASTER DUB** is then prepared and placed in the vault.

The master is then sent for **DUPLICATION** of release copies. This process normally takes 5 to 10 working days for quantities of 300. A "check print" is reviewed and a quality control check is conducted.

The release copies are then sent to the **DISTRIBUTION** center.

**Typical Video Production**
(Percent of Effort)

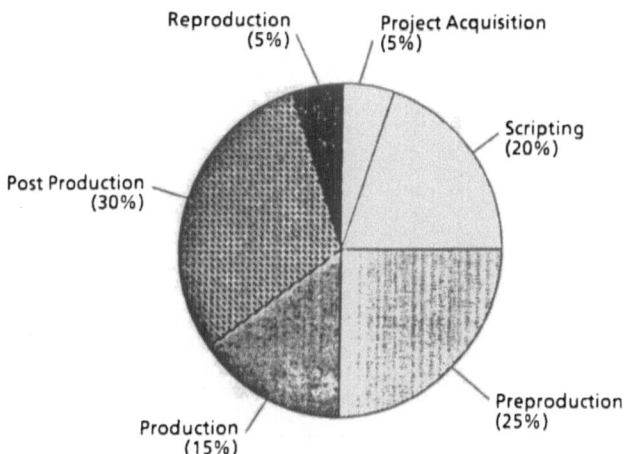

Reproduction
(5%)

Project Acquisition
(5%)

Scripting
(20%)

Post Production
(30%)

Preproduction
(25%)

Production
(15%)

Because each production is unique, it is impossible to give firm estimates of the time required for each phase. The complexity of the project, the number of scenes, locations, set-ups, and special requirements, determine the amount of effort. Client commitment is another important factor. Is the client available and are reviews conducted promptly?

Typically, scripting requires three to six weeks; preproduction two to four weeks; production a few days to two weeks; post production three weeks; and, reproduction one week. These time frames exclude client reviews. This is not to say all productions require this much time. The *Automated Enrollment System* was scripted and produced in two weeks; but, this was an exception

The following is a brief explanation of the video production process. Points requiring client interface are indicated by an asterisk. There are as many as 16 times when client interface is essential.

3

(Courtesy of the U.S. Postal Service.)

Joye R. Swain
Address
City, State   Zip
Telephone Number

Treatment

J.D. McCarty Center for Handicapped Children--

A Place with a Purpose

(Working Title)

by

Joye R. Swain

Our goal in this video is to show that the McCarty
Center prepares people with cerebral palsy and/or neurological
damage to function to the best of their capabilities--this is the
purpose of the Center.  Consequently we will emphasize
accomplishments of children.

Many people are repelled by handicapped children and do
not want to look at them.  In order to catch our audience's
attention without turning them off, we will open at the swimming
pool.  The handicaps will not be immediately obvious.  We will

(Courtesy of the J.D. McCarty Center for Handicapped Children.)

focus on Marty as she swims--then follow her as she painfully and with great effort exits the pool by herself and gets into her wheelchair.

We follow her to her apartment where a party is in progress.

Titles over party scenes featuring other handicapped children.

This is a going away party--a time of great pride and happiness. The setting is an apartment where teenagers practice living on their own. Marty has been living here and she can now take care of herself. This is her last day at the Center. Tomorrow she will move into her own apartment.

Then we flash back to show stages of treatment which enable children to reach this stage.

The middle of the video shows scenes which are typical of what might happen in families with a developmentally handicapped child. We see a typical session in the hospital with a doctor giving the news to the new parents. A social worker helps them cope with the diagnosis and offers information about the McCarty Center.

We see a child arrive at the Center for the first time and we watch a typical evaluation session. This middle portion of the video will show a wide range of types of impairment treated at the McCarty Center and a wide range of the types of treatment available--both in-house and for outpatients.

We watch physical therapy, occupational therapy, speech

J.D. McCarty Treatment--3

therapy.  We see special equipment used for support and communication.

We see the dentist and children recovering from specialized operations common for these children.

We discuss schooling and recreation.

Counselling services are shown via a counselling session with the emphasis being upon normal problems which these children share with others.

We see a montage of different religious services available to children.

Interaction between parents and staff and love and kindness are emphasized along with the treatment itself.

We close with a return to Marty.  The message is that not all children can attain her level of capability, but the Center strives to help each child do as well as possible.  For some, like Marty, that means being able to live independently.

We wind up with mixed shots of Marty at her party.  The focus shifts to another teenager who seems to show unusual attention to the apartment.  This is Jason, who will move into this supervised apartment as soon as Marty leaves.  He can hardly wait.

Close on the excited face of Jason as he rubs his hand over a sofa or such.

Titles at the end give addresses and telephone numbers for people who might be interested in these services.

This is to be an upbeat video emphasizing

J.D. McCarty Treatment--4

accomplishment, but it is not to be unrealistic and to promise
that all children can become independent.  Shots will be realistic
and show children struggling to walk and feed themselves and
communicate.  But there will also be shots of children smiling,
playing, enjoying themselves.  The emphasis is on a can-do spirit
of love and kindness.  We want our audience to like these children
and to understand their struggles and not to be so disturbed by
their handicaps that they withdraw from the presentation.


The End


Approved for the McCarty Center _____

FADE IN:

INT.-DAY-SWIMMING POOL

CU, MARTY in pool. You cannot
tell she is handicapped.

PULL BACK to show others in
pool. Now it becomes clear
that this is a group of
handicapped children.

PAN TO follow Marty as she
exits the pool, puts on
braces, and pulls herself into
a wheelchair. As she rolls out
of the shot she waves good by
to the others.
                    CUT TO:

EXT. - DAY - MARTY'S APARTMENT
MS, entrance to Marty's
apartment. Marty rolls up.
                    CUT TO:

SOUNDS:  Pool sounds.  Children
laughing.

MOTHER:  This is my daughter,
Marty.
I know that every father thinks
his daughter is special, but
Marty
is special to a lot of people
...not just me.

Today is a special day for
Marty.  It's her last day at
the McCarty Center for
Handicapped Children.  Tomorrow
Marty moves into her own
apartment.  She already has a
job in a sheltered workshop.
So, she's on her way to being
an independent adult.

(Courtesy of the J.D. McCarty Center for Handicapped Children.)

McCarty--2

CU, Marty opens the door.

CUT TO:

INT. - DAY - MARTY'S APARTMENT

MS, living room.  It is full        MIXED VOICES:  Surprise!  We'll
of balloons, banners, and           miss you, Marty.  Wow, it must
handicapped children, teens, and    be great to have your own
McCarty staff.                      apartment.  I can't wait to get
JASON is prominent.        .        this apartment.
* TITLES OVER *40 + 8 association*  Congratulations!
*present 12*
"J.D. MC CARTY CENTER FOR           SOUNDS:  Party sounds.  Music.
HANDICAPPED CHILDREN--              SOUNDS DOWN AND OUT.
A Place with a Purpose"             NAR:  But life wasn't always so
                                    full of hope for Marty and her
                                    parents.

                 DISSOLVE:                         *Rewrite*

CU, parents hugging OWEN's BABY     NAR:  It was, at the hospital that
who could be baby Marty.            the Doctor
                                    ~~social worker who~~ first told
                                    them about the J.D. McCarty
                                    Center for Handicapped
                                    Children.

MCU, above to include brothers      But she also emphasized how
and sisters, grandparents, and      important early bonding is--
such.                               that time when the family and
                                    child form the links essential

United   States   Postal   Service
Integrated   Logistics   Support   System

Introductory   Videotape   Script

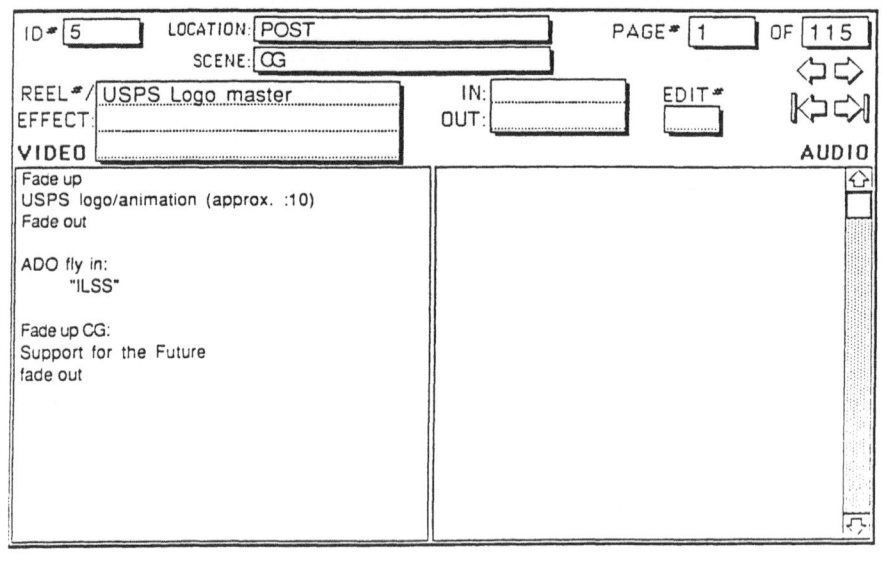

(Courtesy of the U.S. Postal Service.)

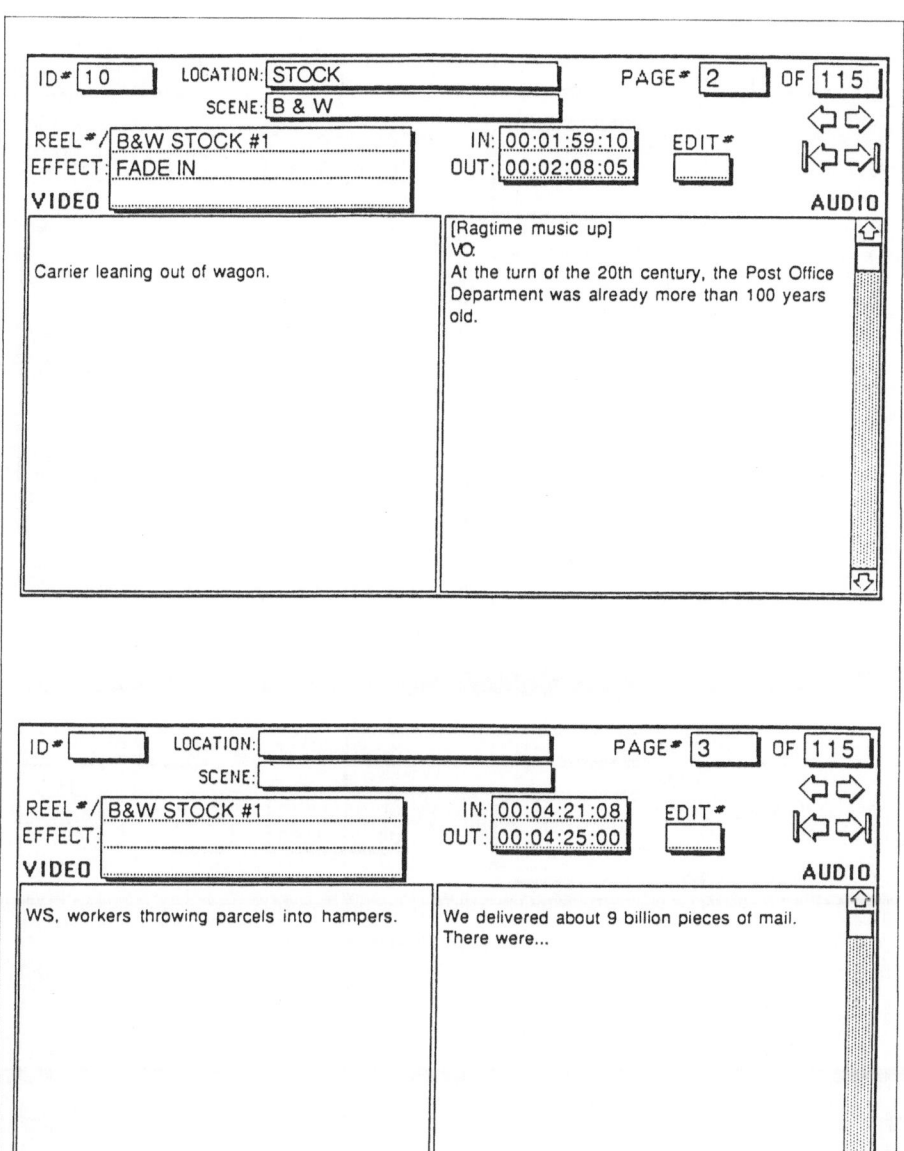

ID# 10    LOCATION: STOCK         PAGE# 2   OF 115
          SCENE: B & W
REEL#/ B&W STOCK #1         IN: 00:01:59:10    EDIT#
EFFECT: FADE IN            OUT: 00:02:08:05
VIDEO                                              AUDIO

Carrier leaning out of wagon.

[Ragtime music up]
VO:
At the turn of the 20th century, the Post Office Department was already more than 100 years old.

---

ID#         LOCATION:              PAGE# 3   OF 115
            SCENE:
REEL#/ B&W STOCK #1         IN: 00:04:21:08    EDIT#
EFFECT:                    OUT: 00:04:25:00
VIDEO                                              AUDIO

WS, workers throwing parcels into hampers.

We delivered about 9 billion pieces of mail. There were...

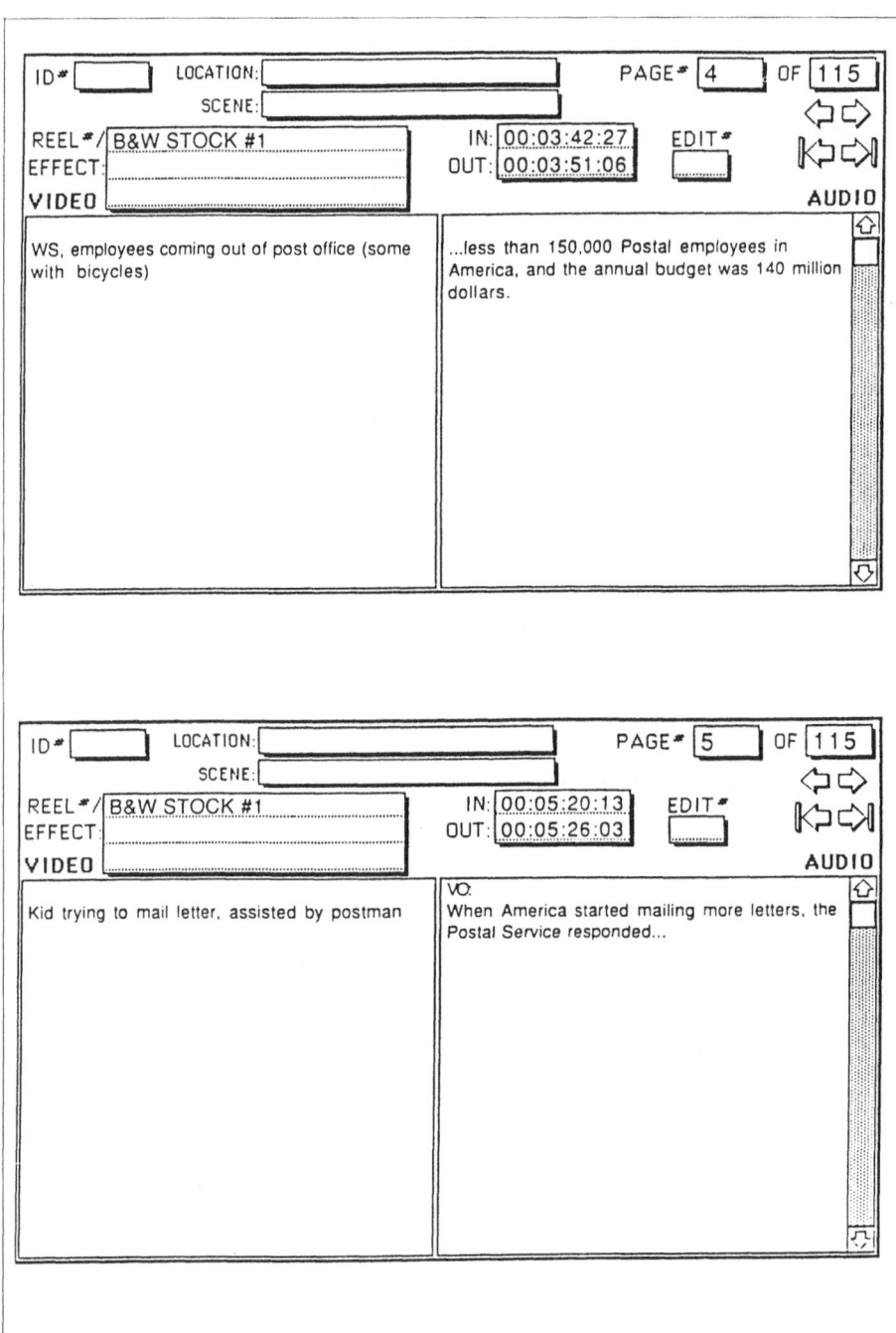

ID#　LOCATION:　PAGE# 4　OF 115
SCENE:
REEL#/ B&W STOCK #1　IN: 00:03:42:27　EDIT#
EFFECT:　OUT: 00:03:51:06
VIDEO　AUDIO

WS, employees coming out of post office (some with bicycles)

...less than 150,000 Postal employees in America, and the annual budget was 140 million dollars.

ID#　LOCATION:　PAGE# 5　OF 115
SCENE:
REEL#/ B&W STOCK #1　IN: 00:05:20:13　EDIT#
EFFECT:　OUT: 00:05:26:03
VIDEO　AUDIO

Kid trying to mail letter, assisted by postman

VO:
When America started mailing more letters, the Postal Service responded...

# The Interactives

One of the most active, promising, and booming fields in audiovisual today is that of interactive programs. As a scriptwriter, this is one of the audiovisual fields in which you may find work.

What's an interactive program?

Break it down. *Interactive* means a system in which a person performs some act in order to get a desired response from the system. The automated banking machine offers a familiar example. To make a withdrawal, you punch an activating button. An instruction then flashes on the screen. It asks that you punch in a personal number that establishes that you have an account with the facility. Another message then appears, this one asking how much you want to withdraw. The machine thereupon determines whether your account is large enough to permit such a withdrawal. If it is, the machine delivers the money to you.

A self-service elevator is another interactive machine. From an array of options, you select the floor at which you want the elevator to stop and push the button indicating that particular floor.

In other words, the user influences the machine's behavior. In the case of the bank's automated teller, only one user at a time is involved. The elevator, in contrast, can respond to more than one person at the same time.

An *interactive audiovisual program* operates on the same principle. The viewer to some degree controls the sound or pictures presented to attain a desired end—win a game, learn a lesson, determine a course of action, or the like.

"Anything that lends itself to a lot of decision making is a good candidate for interactive," says Michael Spencer, Executive Producer for the Center for Instructional Development and Evaluation of the University of Maryland.

To make such decision making possible, the audiovisual program utilizes what's known as *branching*—that is, it offers two or more possible courses of action between which the viewer must choose. In a game, this may mean picking the proper road to follow in order to win. In a teaching program, it could be selecting the correct answer to a problem in addition. If the viewer makes the right decision, he or she is permitted to tackle the next step toward winning the game or finishing the exercise. If the choice is wrong, the viewer loses a turn, or has to view the segment on addition again.

Most such programs are presented by means of interactive *videotape,* interactive *videodiscs,* or *computer-based programming.* (Incidentally, the audiovisual world prefers to spell the term *disc.* Computer types ordinarily spell it *disk.*)

Interactive videotape was an early development, but it had a problem. For maximum effectiveness, it's necessary in any interactive program to jump from one segment to another, backward or forward, on the basis of the choices the viewer makes. But tape is linear. It's set up with the shots in sequential order. For it to move backward or forward to reach the correct branching point takes time; therefore, it's slow. It's used primarily for programs involving only starting and stopping commands.

An interactive videodisc, in contrast, is nonlinear. It provides computer-controlled random access to the stored material. That is, in effect it can jump backward or forward from frame to frame, thus making possible almost instantaneous development of the program.

Because of its greater speed, interactive videodisc has replaced interactive videotape to a large degree in most circumstances.

The standard 12-inch laser videodisc offers the capacity of a mix of 54,000 still frames, 30 minutes of full-speed motion, or 30 minutes of audio per side, with exceptional picture quality and full-bodied, high-quality sound.

The third interactive alternative is computer-based programming. Elements include computer graphics, video from a videodisc, and audio. Computer-based training (CBT) uses a computer to present learning situations. The viewer interacts with the program, and the program can be modified while in use.

A high proportion of videodisc presentations rely heavily on computer software, which is prepared by a computer programmer. Using this software is termed *authoring.*

Put in the simplest possible terms, authoring is an approach that makes it possible for people who lack computer programming training to prepare programs for computer- or videodisc-based systems. In effect, it provides a format into which program material can be fitted.

In the past, a person had to know how to do computer programming to prepare computer-based training. Now, with all of the authoring systems and computer-generated graphics available, it's possible for a writer, using the computer only as a writing tool, to prepare computer-based programming for training, educational, or entertainment material. An *authoring system* or *language* makes it possible for people whose skill is pretty much limited to typing to produce interactive programs. So the field is now wide open and growing.

Up to a point, that is. Actually, it's almost impossible for anyone without at least a rudimentary knowledge of computer procedures to go far in this field. But more of that later.

## INTERACTIVE TAPES AND DISCS

How does an interactive videotape or disc program come into being—the content, not the technology?

The first step is taken by the sponsor, who decides on the program's subject and its goal—to teach children to write or to train hospital emergency room personnel, for example.

That out of the way, at least three people ordinarily are involved in creating an interactive program: a *content specialist*, an *instructional designer*, and a *scriptwriter*.

The content specialist is a subject matter expert. He or she provides the instructional designer with whatever information is needed about the program's topic.

What's the instructional designer's role in all this?

1. To decide what the program is to include.
2. To select the branching points—that is, the spots where the viewer demonstrates an understanding of the key facts by making a correct choice between alternatives.
3. To specify what information—possibly which shots—is to be included in each unit.

To this end, the instructional designer must be an expert in learning theory and viewer psychology. He or she plans how to translate the information on the topic into a presentation that will convey the message to the audience effectively.

The scriptwriter, in cooperation with the designer, sets down this presentation. Together, they may decide what images go onto the screen, the camera angles, the length of time an image stays on the screen, the order of sequences, and the like in whatever form the designer specifies.

Thus, the writer can't function merely as a writer in an interactive program. Besides mastering scripting skills, the writer also must frequently master those of the designer. So, of necessity, the writer becomes a "hyphenate"—a designer-writer, most often.

This team begins to prepare a program. The basic decisions on content made, the instructional designer, alone or in collaboration with the content specialist, works out a *flow chart* of the project.

A good example of this is material developed for the Essex Corporation's Argus Project. Illustrated are the Project's basic plan, examples from

ARGUS PROJECT INTERACTIVE VIDEODISC
Compact Laser Designator

This interactive videodisc was commissioned to accomplish several purposes. Primarily, it was a prototype designed to demonstrate potential for using videodiscs in support of Naval Special Warfare (NSW) training. This prototype was developed using the Compact Laser Designator as a candidate system for the evaluation. In addition, this videodisc was designed to demonstrate the manipulative power and versatility of the computer-managed videodisc format.

This program was developed on a Sony View System using "C" as the computer authoring language. The "View" system is configured similar to a PC AT with an external 30 meg hard drive, various expansion boards and interfaces. The videodisc player is a Sony LDP 2000 built into the View system. A Sony color monitor is a PVM - 1271Q. A mouse is used as the primary means for program manipulation.

This is a menu driven videodisc with three main separate categories: Training; Pre-Mission Test; Mission. Each of the segments have extensive sub menu avenues and options. The Training Segment deals with technical aspects of orientation, familiarization and operation of the Compact Laser Designator Unit. Sub-menus provide options for the user to view, obtain particular detail, and closely inspect all physical aspects of the unit. Short manipulative exercises are imbedded as a means to measure user's ability to comprehend technical aspects.

The Training Segment is designed as a preliminary introduction for the Pre-Mission Test Segment. In the Pre-mission Test Segment, the user must apply information from the Training Segment to demonstrate proficiency in practice lasing different target configurations. Users must achieve minimum acceptable scores in order to gain access to the third segment - the Mission. Should users fail to reach 80% or better scores, they are denied access to the Mission Segment until they do reach the acceptable levels. Remediation and practice is available to assist the user to reach acceptable scores.

The first two segments are technically oriented whereas the third segment, The Mission, is related to tactical decision making. Having successfully completed proper training in the use of the CLD, the user is provided the opportunity of a simulated mission to exercise skills and training. The objective of The Mission is to successfully infiltrate a SEAL Team element into a country called La Mancha for a joint procedure for laser desig-nation of a particular target for aerial delivered laser-guided weapons. During the mission, the user is faced with decision-making situations which branch into various scenarios based upon individual selections. Exercises in navigation, infiltration, team leading, patrolling, locating and lasing a target, coor-dinating with strike aircraft to direct "smart" bombs to the

(Courtesy of the Essex Corporation, 1430 Springhill Road, McLean, VA 22102.)

target are incorporated.  Enemy encounters,  and consequences are
designed to  provide "simulation" experience for trainees.  There
are a number of different avenues and options  available to reach
the specific target.

At the  conclusion of the Mission, the user is provided with
an assessment of his  decision-making activities  with an evaluation of  each decision.   Recommended  and correct selections are
emphasized for reinforcement purposes.

Graphic overlays, motion and still frame  combinations, plus
animated  sequences  are  utilized  extensively to enhance visual
interest and stimulate user  participation.    Every  attempt has
been made  to provide  a visually appealing, comprehensive, fully
interactive,  successful  training  demonstration,  not  a  fully
interruptive distracting presentation.

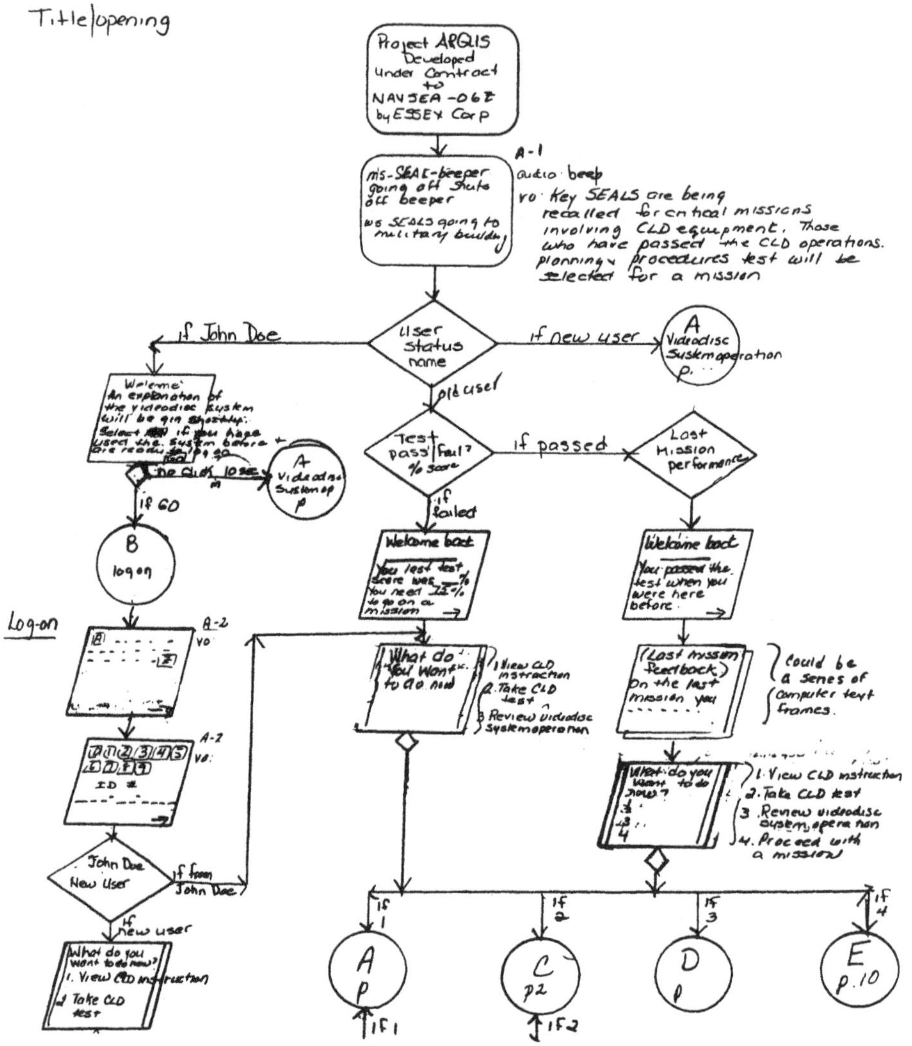
Title/opening
Project ARQIS Developed Under Contract to NAVSEA-06 by ESSEX Corp
A-1
audio beep
vo: Key SEALS are being recalled for critical missions involving CLD equipment. Those who have passed the CLD operations planning & procedures test will be selected for a mission
User status name
if John Doe
if new user
Videodisc System operation
old user
Test pass/fail? % score
if passed
Last Mission performance
if failed
Welcome back
Welcome back
You passed the test when you were here before
B log on
if GO
Log-on
A-2
VO
ID #
John Doe New User
if from John Doe
if new user
What do you want to do now?
1. View CLD instruction
2. Take CLD test
3. Review videodisc system operation
(Last mission feedback) On the last mission you
Could be a series of computer text frames.
What do you want to do now?
1. View CLD instruction
2. Take CLD test
3. Review videodisc system operation
4. Proceed with a mission
if 1
if 2
if 3
if 4
A p
C p2
D p
E p.10
IF 1
IF 2

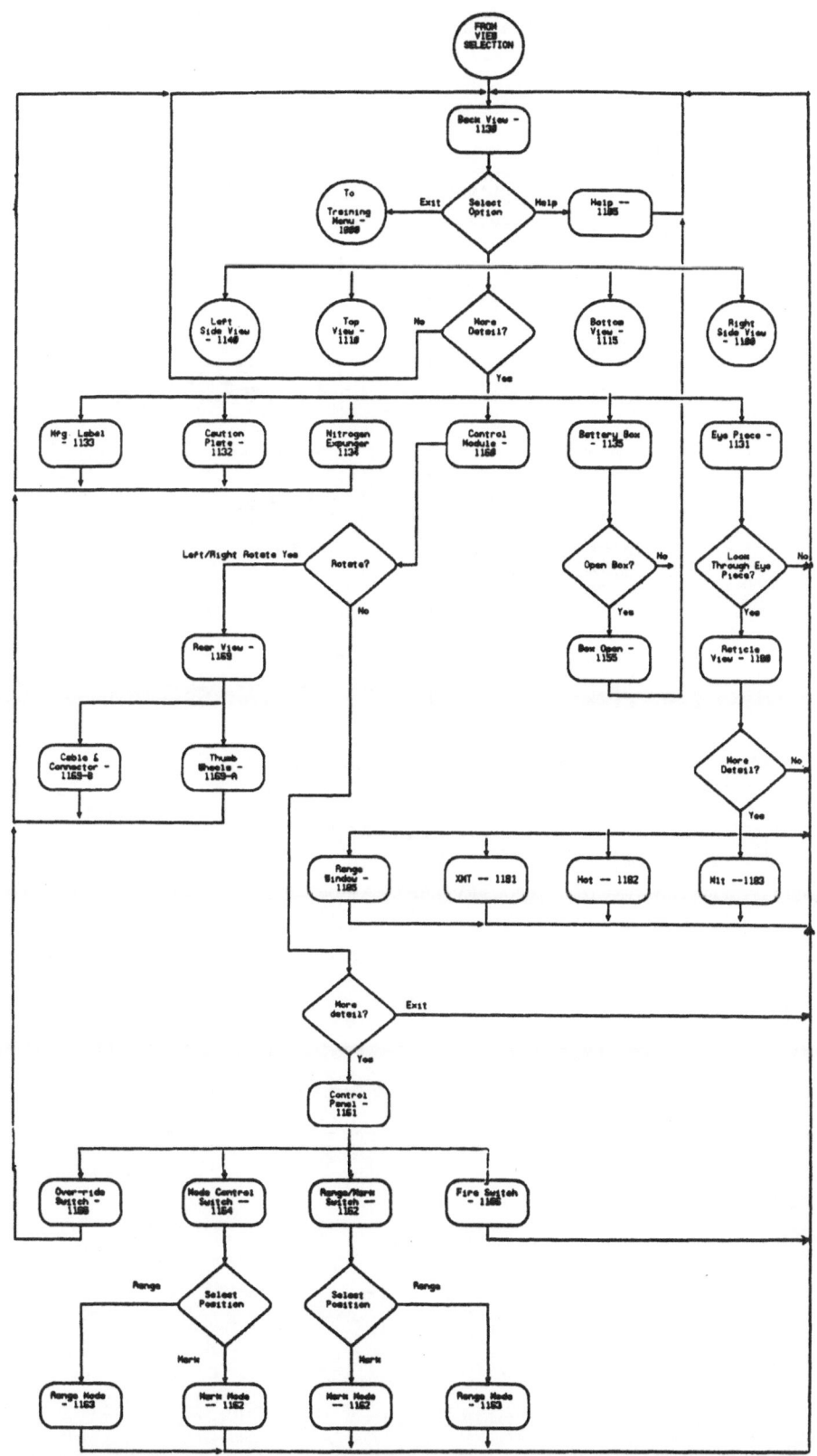

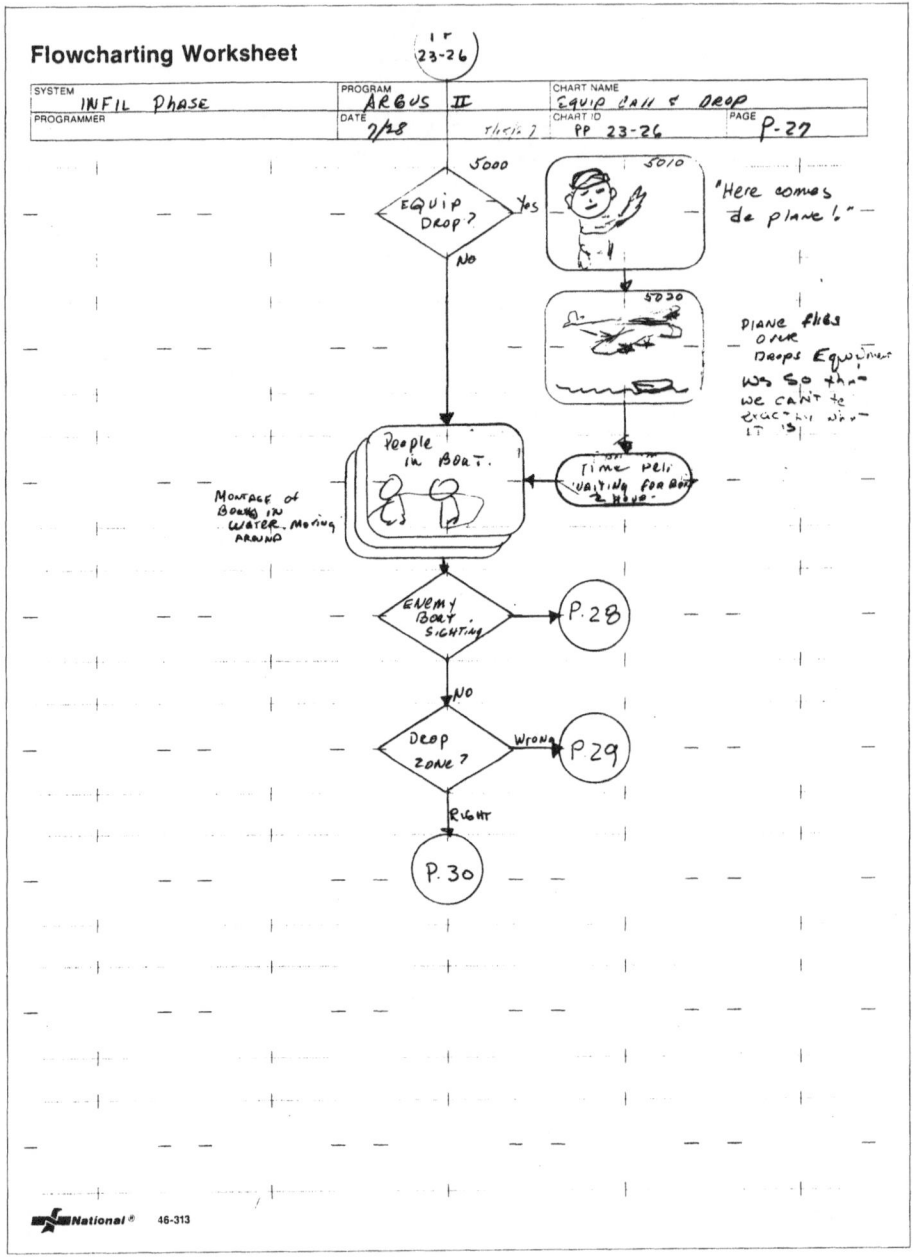

**Flowcharting Worksheet**   (23-26)

| SYSTEM | PROGRAM | CHART NAME |
|---|---|---|
| INFIL Phase | ARGUS II | EQUIP CALL & DROP |

| PROGRAMMER | DATE | | CHART ID | PAGE |
|---|---|---|---|---|
| | 2/28 | Thesis 7 | PP 23-26 | P-27 |

5000 — EQUIP DROP? — Yes

5010 — "Here comes de plane!"

No

5020 — Plane flies over Drops Equipment was so that we can't trac~ by whe~ IT 's|

People in Boat.

MONTAGE of BOATS IN WATER. Moving AROUND

Time peli waiting for Boat 2 hous.

ENEMY BOAT SIGHTING → P.28

No

Drop Zone? — Wrong → P.29

Right

P.30

National ®   46-313

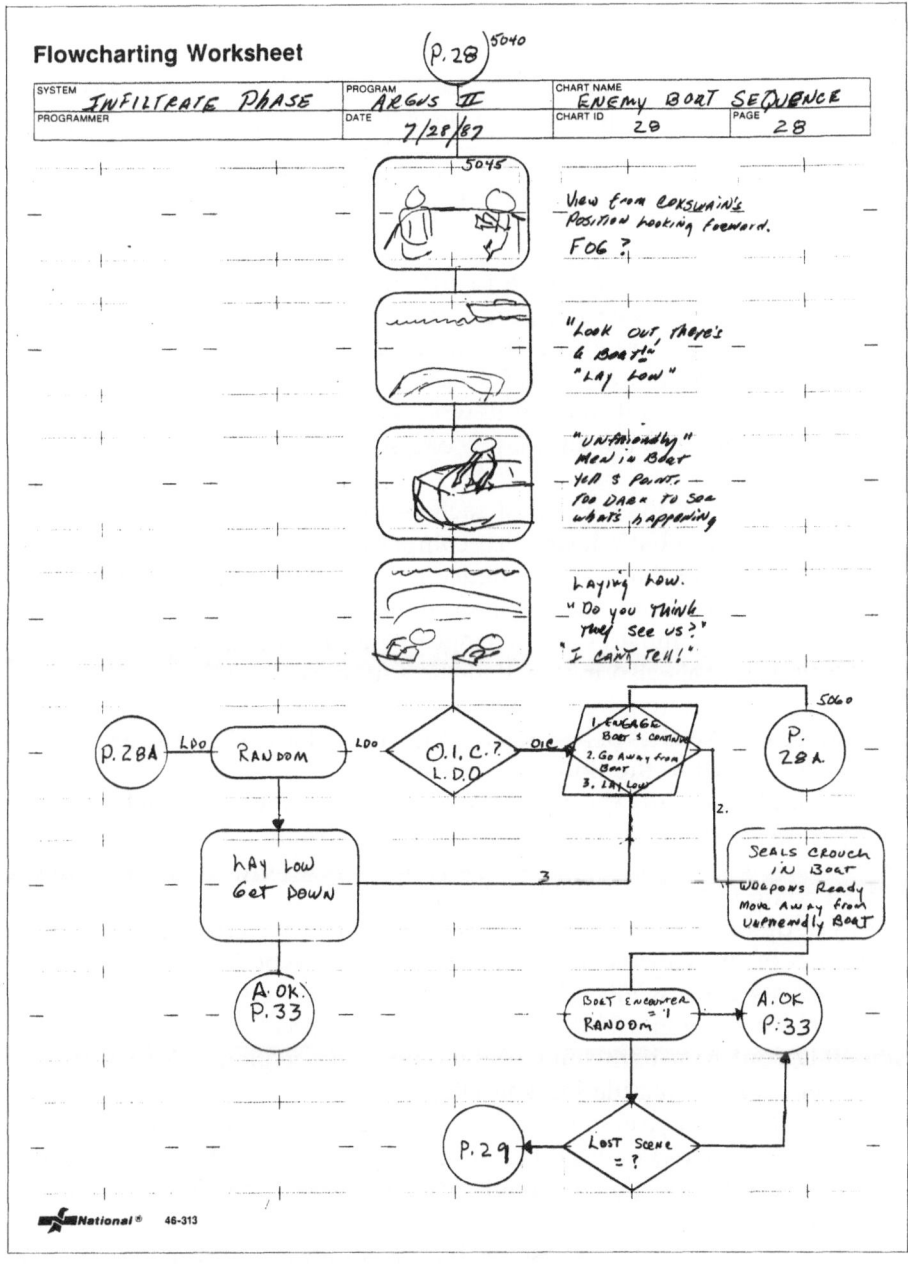

**Flowcharting Worksheet**   (P. 28) 5040

| SYSTEM *INFILTRATE PHASE* | PROGRAM *ARGUS II* | CHART NAME *ENEMY BOAT SEQUENCE* |
|---|---|---|
| PROGRAMMER | DATE 7/28/87 | CHART ID 28 / PAGE 28 |

5045

View from Coxswain's Position looking foreword.
FOG ?

"Look out, there's a Boat!"
"Lay low"

"Unfriendly" Men in Boat — YOA 3 Points —
Too Dark to see whats happening

Laying Low.
"Do you think they see us?"
"I can't tell!"

5060

P.28A —LDO→ RANDOM —LDO→ O.I.C.?  L.D.O —OIC→  1. ENGAGE Boat's contents  2. Go Away from Boat  3. Lay Low   →  P. 28A

LAY LOW  Get Down    ←3

2.

SEALS crouch in Boat — Weapons Ready Move Away from Unfriendly Boat

A.OK. P. 33    BOAT ENCOUNTER = 1 RANDOM    A.OK P. 33

P. 29 ←  Lost Scene = ?

National®   46-313

the initial flow chart worksheet, the completed flow chart, and individual frames.

A flow chart is a sort of diagram that breaks down the program's predetermined goal, its lesson, into steps. Each step is a sequence, a group of related shots, and each calls for a decision on the part of the user as to the best answer between possible explicit choices offered on a film or video frame. The user registers this choice by pressing a button, touching the screen, or the like.

The program then moves on to another sequence—one if the answer is right, another if it is wrong. And so on, sequence by sequence, until the user reaches the program's desired training goal.

The flow chart pinpoints the possible routes a user may take through the program to the end, nailing down the decision point to be presented in each sequence and the alternative courses to be offered in consequence of each decision.

Along the line, the content of each sequence must be decided upon—what's seen and what's heard. To complicate the problem, the designer must bear in mind that each sequence must be so designed that it can be switched to one or more additional positions within the program as needed. In other words, it must be interchangeable upon demand.

Ordinarily, this also means that the soundtrack will be recorded separately so that any required switching need not be limited by it.

Notice the sketches of key details in the flow chart. This is an adaptation of the storyboard to the flow chart format.

The main thing the scriptwriter needs to remember is that while everything in the program needs to hang together, visually each sequence must be so designed that it can stand alone, since you don't necessarily know what's going to be juxtaposed with what.

In addition, the designer has to work with the scriptwriter to make sure each of the viewed segments fits together. Whether you see two of these segments or twenty, they must make sense, regardless of the order in which you see them. Thus, in linear presentations, the shots tie together smoothly. But in a branching presentation one sequence may feature a character acting or simulating speech.

Since this sequence may be used six times in assorted situations, continuity must be loose enough that a "talking head" or what have you can be dropped into a sequence about road building, a machine shop, or an office meeting.

Thus, the designer must establish (1) what idea or information is to be conveyed to the viewer, (2) what sequences are to be selected, and (3) in what order they are to be arranged.

Let's say that the program is a game in which the viewer strives to move from an entry point on the board to an exit. To do this, he or she must make correct decisions as to the road to follow at five different intersections.

Or assume that the project is a comprehensive truck repair program.

The viewer must choose which section he or she wants to learn about: suspension, transmission, ignition, brakes, or whatever.

The viewer does this by pressing a button—for the brakes unit, let's say. Whereupon, the program jumps to that portion, ignoring all the rest.

Within that section, the viewer wants instruction on brake shoe installation. He or she pushes the "brake shoe installation" directory line. The program jumps to that unit, presenting it in terms of questions to be answered and decisions to be made. And so on.

As far as the formats of flow charts and storyboards are concerned, one of the problems is that so far no standardization has been established. Not even the symbols used in presentation are universal. But they do share an idea: The display is designed to tell the client or the production team how the ideas are to be presented.

In consequence of this lack of standardization, the writer must learn the particular "language" in which each program is to be offered to the viewer. Experience is the only teacher, and there's little point to discussing procedures in detail.

It's at this juncture that the scriptwriter comes in. Following the directions of the instructional designer, the writer sets down the shot descriptions, the narration and dialogue lines to be spoken, or both.

"The writer is really just a clean-up editor," one expert says. Another adds, "A lot of people do scriptwriting or other kinds of writing, but they really are designers."

Actually, the scriptwriter's basic skills in interactive are the same as they are in any phase of scripting, but broken down into small segments or sequences. That is, the writer writes shot-by-shot instructions for the production crew, just as in any audiovisual script. But he or she must think in nonlinear terms rather than linear ones. That is, instead of working on the assumption that one shot will necessarily follow another in a predetermined order, the writer must consider each as a separate-but-related entity that may be used over and over, pulled out of its original context and inserted elsewhere in order to carry out the program's mission.

This leap from linear to nonlinear calls for a whole new mode of thinking on the writer's part, and it's hard to make, as witness an observation by Thomas H. Bair of Kinton, Inc.:

> Probably the biggest problem of working with scriptwriters is teaching them to think disjointedly, helping them to work in terms of segments that may or may not be seen together, and trying to understand the various possible ways that the program might flow depending upon how someone might use it. The viewer has never seen a particular scene because with the choices that were made they've gone from one to another one. But the scene still has to be developed.

Can you make this transition? That depends on you. Given perseverance and a degree of insight into what you're trying to accomplish, it's certainly possible, for others have done it. You'll see what's involved shortly.

How do you prepare yourself to be a scriptwriter of interactive presentations? Michael Spencer has some practical ideas on the subject:

> It seems to me that you're going to need to sit down and work with—and play with—interactive videodiscs, a variety of programs, plus spending quite a bit of time with a videodisc production team.
>
> You should get a job—any kind of job—on a production team because I think it's very critical that someone be able to understand what this sort of thing means in the overall flow of the program. They should be able to understand the flow chart, how it integrates to the final program, and be able to abstract at this level and be able to understand how these specifics are going to be translated from this abstract level. Otherwise you'll see the kind of things I've seen happen with video production facilities that decide that they're going to go into interactive videodisc. They produce beautiful video, but it's oftentimes not good interactive videodisc. They don't use the capabilities. They do something that's much more linear than it needs to be or should be. They're still trying to tell a linear story. Interactive's a different way of looking at communicating a message, when you think of it in these discrete chunks that can be put together in various ways.

A major point of confusion for new writers is the fact that, because interactive programs branch repeatedly and disc space is limited, a given picture may be used several times, regardless of whether the picture is still or motion. Further, because of this, multiple situations may call for different comments.

This brings forth a very pointed word of warning regarding normal linear writing versus interactive video, from Jack Hirschfield, a producer for the U.S. Postal Service's Media Branch:

> A person who likes to write should probably not be involved in interactive video. Because the impulse to write is different from what is needed in interactive. Those that like working with language and marrying the language to visual stimulation towards meaning don't belong in interactive video. While it's true that at some point decisions will have to be made as to what words people will hear and the person who's trained to write for the ear is important, he's got to think of the screen as a visual thing.

Interactive video is broken down into three levels of complexity. Level 1 is quite simple. The only available controls are play, stop, fast forward, and reverse. A videocassette player can be operated on this level.

Level 2 uses a videodisc player that has a built-in microprocessor. The microprocessor reads the program off the videodisc. The viewer punches precoded buttons to instruct the player to search particular parts of the program. This level is used frequently in commercial settings when stand-alone operation is preferred or necessary. In a stand-alone setting, a videodisc player is usually housed in a kiosk.

Level 3 is characterized by a videodisc player that is controlled by an external computer. The computer controls all operations of the videodisc player. This system offers the greatest potential for future development.

Now, let's look at examples of interactive videodisc- and computer-based training. The first videodisc example is taken from the U.S. Department of the Interior's Yosemite #1 program. A simple script, it shows how video and audio are combined to make an effective and impressive presentation at low cost.

The Park Service's first ventures into the video field used videotape players, but found them expensive and subject to too much downtime.

"Most of our video installations are visitor-activated," explains Thomas A. Kleiman, the Service's Audio Production Officer. "In 1980, we purchased our first videodisc players. Extensive field testing showed us that they were dependable and reliable. By pushing the button of his choice, the visitor gets the information he's interested in."

The second videodisc example is a fragment from a rather elaborate air control script prepared by the Center for Instructional Development and Evaluation of the University of Maryland. The Hampton Project flow chart is followed by a piece of the actual script.

The Park Service presentation, you'll note, is prepared without a flow chart because none is needed. While the video is interactive, the information offered is straightforward and no organized learning plan is necessary. The Hampton Project, in contrast, calls for a thoughtful evaluation of the possibilities, so the development is carefully laid out.

Where the script is concerned, the Park Service relies on a description of simple pictures. The Hampton Project uses rather elaborate video illustrations. Note that each Hampton picture stands more or less independently. Consequently it can be switched from position to position.

The third example is the complete master script of an interactive touchscreen laser disc written by Ben Walker. Titled "The Problem with Shirley," this program's purpose is to help supervisors improve their handling of problems with employees.

The flow chart traces the course of the program's development, laying down the points at which it branches, corresponding to possible supervisor reactions, desirable or undesirable, to Shirley's behavior. The program's premise—that is, the topic and conclusion to be presented—is nailed down at the top of the chart: "In any confrontation, the first few moments are crucial."

The development of the program follows, with each frame in a circle and numbered (1 to 16) and possible responses to each assigned a letter (A to C).

The actual script sets forth, frame by frame, the visual element on the left and the sound on the right. An obviously angry Shirley charges into her supervisor's office. Three possible supervisor responses to her outburst bring branching to frames 3, 4, or 2.

```
                                        YOSE-TV-64
                                        5/1/84, Revised 5/8/84,
                                        7/6/84, 5/6/85,
                                        Narrated 7/2/85




                    YOSEMITE #1

                    WINTER VERSION



        SEQUENCE #1


        FADE IN ON EARLY MORNING        SFX TO MATCH PICTURE
        MONTAGE OF YOSEMITE VALLEY                               (:05)
        IN WINTER.  MISTY MEADOW DEW
        ON GRASS.  LONE JOGGER ON
        LONG LENS.




                                        V.O. NARRATOR:

                                        If you have only one day to see Yosemite

                                        here are a few suggestions for using your

                                        time to best advantage.




                                        PAGE TIME:

                                        RUNNING TIME:
```

(Courtesy of the U.S. Department of the Interior, National Park Service, Division of Audiovisual Arts.)

```
                                        YOSE-TV-64
                                        5/1/84, Revised 5/8/84,
                                        7/6/84, 5/6/85,
                                        Narrated 7/2/85

SEQUENCE #2
                                        V.O. NARRATOR:

INTERIOR OF THE VISITOR CENTER.         The exhibits and rangers in the Visitor

SEVERAL VISITORS ARE STANDING           Center can help you get started.

AROUND RELIEF MAP WITH RANGER.

MS OF GROUP.



                                        RANGER'S VOICE:

                                        We're right here at Yosemite Valley

                                        Visitor Center.  Outside in back is a

                                        self-guiding nature trail through the

                                        Indian Village and next door is the Indian

                                        Cultural Museum.


CUT TO EXHIBIT                          There you can see exhibits on the rich

PAN EXHIBIT                             material culture of Yosemite's first

                                        people -- the AH-WAHN-EECH-EE




                                        PAGE TIME:

                                        RUNNING TIME:


                                        2
```

7/6/84, 5/6/85,
Narrated 7/2/85

SEQUENCE #3

V.O. NARRATOR:

FADE IN ON SERIES OF CLASSIC       The spectacular scenery of Yosemite Valley
WINTER SCENICS OF YOSEMITE VALLEY
can be enjoyed year-round ... even though

HIGH COUNTRY ROAD.  CLOSED GATE    the high country is now closed to cars.
AND SNOW.

PAGE TIME:

RUNNING TIME:

3

5/1/84, Revised 5/8/84,
7/6/84, 5/6/85,
Narrated 7/2/85

SEQUENCE #4

| | |
|---|---|
| FADE IN ON STATIC SCENIC<br>AS TOUR BUS PULLS INTO SHOT | SYNC SFX OF TOUR GUIDE TALKING ON BUS<br>PA SYSTEM |

V.O. NARRATOR:

CUT TO POV OF VISITOR ON TOUR BUS

If you're here for only a few hours, a

good way to see the Valley in a short time

GUIDE ON TOUR BUS

is the narrated two hour Valley Floor

Tour, offered daily on a fixed schedule.

Information on prices and times is

TOUR BUS PULLING INTO A
PULL OUT

available at the Visitor Center

Information Desk and at lodges.

VISITOR PHOTOGRAPHING

PAGE TIME:

RUNNING TIME:

4

~/~/~~, ~~~~~~~ ~/~/~~,
7/6/84, 5/6/85,
Narrated 7/2/85

SEQUENCE #5

SPLIT SCREEN

V.O. NARRATOR:

CAR DRIVING IN VALLEY

If you have a little more time, you may prefer to drive your own car around the Valley.

ROAD GUIDE

CAR

NUMBERED POST

The Yosemite "Road Guide", keyed to numbered posts along the roads, will help you learn more about the park.

Many of these posts are located at

SENTINEL ROCK

EL CAPITAN

CATHEDRAL ROCKS

BRIDALVEIL FALL

USE BRIEF CUTS WITH CAPTIONS

turnouts that offer impressive views of notable features.

PAGE TIME:

RUNNING TIME:

5

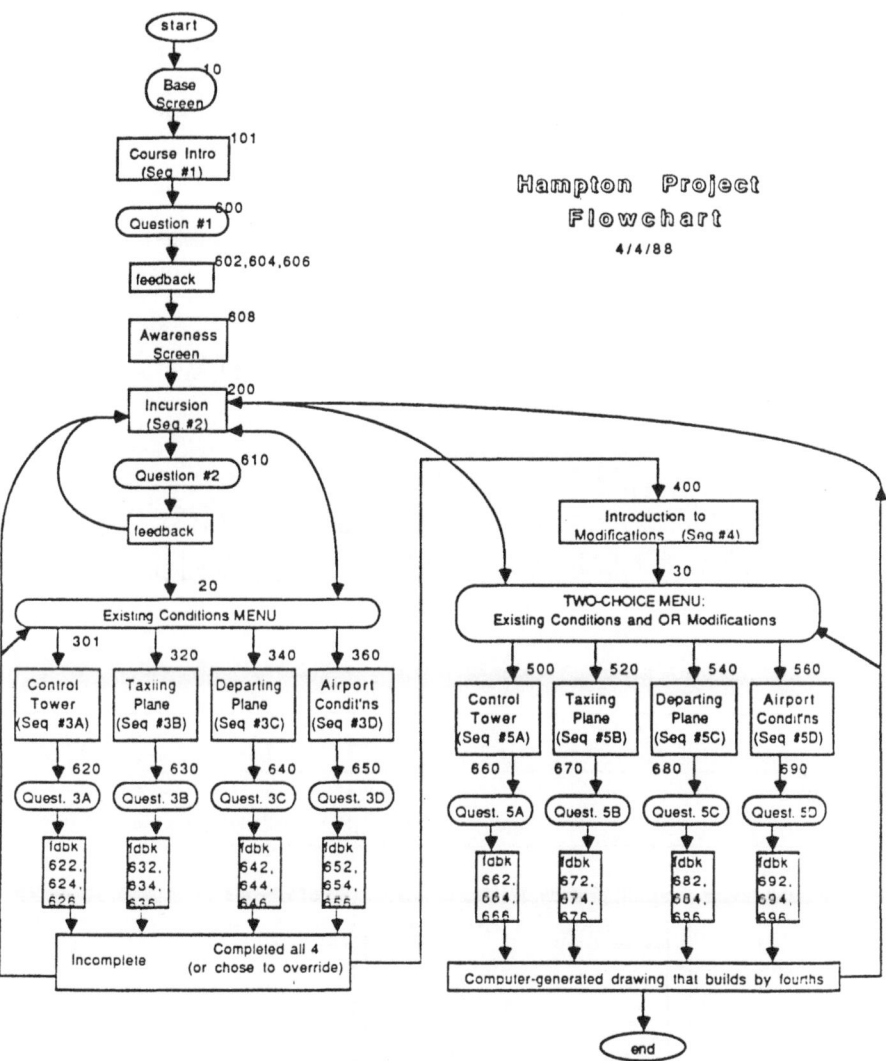

(Courtesy of Herbert B. Armstrong, Coordinator, Airway Science Program, executive coproducer; and Michael H. Spencer, executive coproducer, The Center for Instructional Development and Evaluation, The University of Maryland University College.)

Frame #⌷1 0⌷                                    B A S E   S C R E E N

x

**WHEN THE SYSTEM FAILS:**

*Touch the screen to begin.*

**CONTRIBUTING FACTORS
TO RUNWAY INCURSIONS**

⌷AUDIO⌷

| Condition | Go To | Programming Notes | Display Notes |
|-----------|-------|-------------------|---------------|
| touch | 101 | This base screen is where the user returns to whenever EXIT is selected. | |

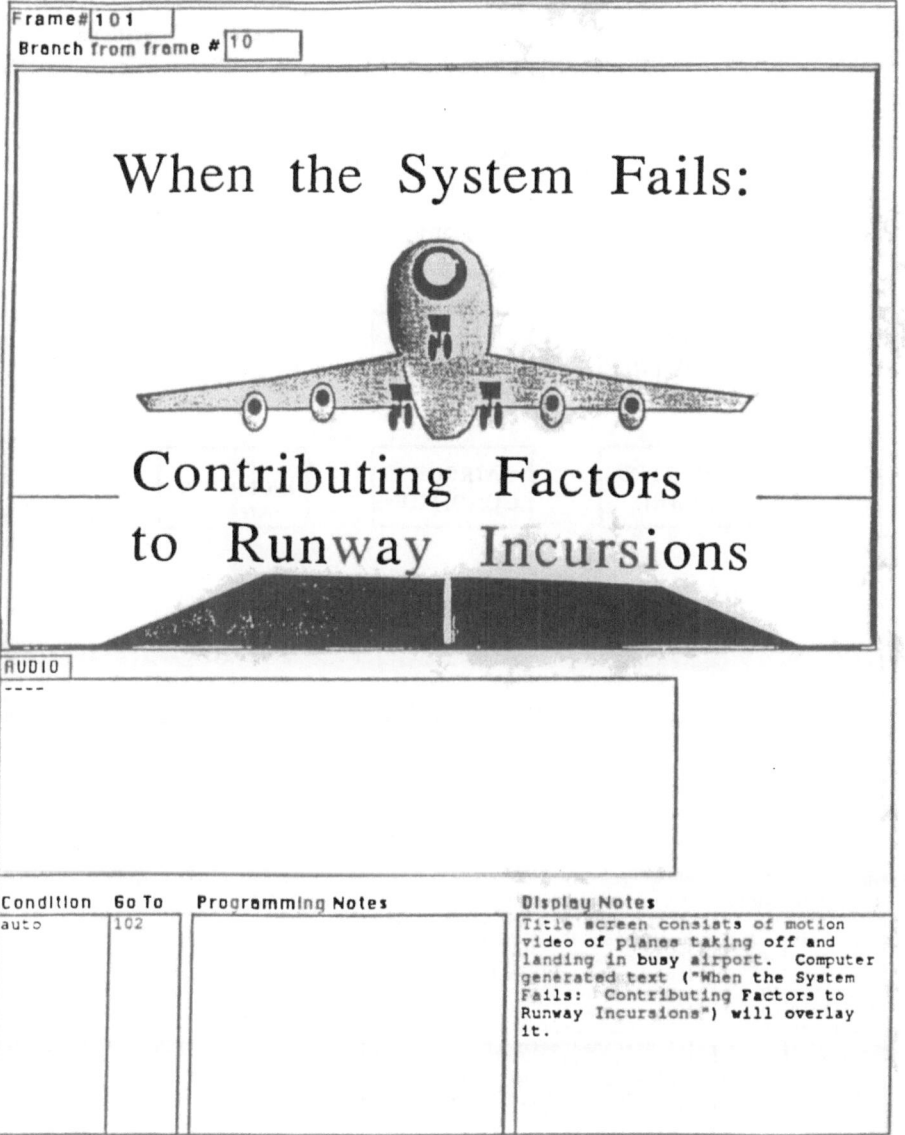

Frame# 101
Branch from frame # 10

# When the System Fails:

## Contributing Factors
## to Runway Incursions

AUDIO
----

| Condition | Go To | Programming Notes | Display Notes |
|---|---|---|---|
| auto | 102 | | Title screen consists of motion video of planes taking off and landing in busy airport. Computer generated text ("When the System Fails: Contributing Factors to Runway Incursions") will overlay it. |

Frame #2 0

x

# MENU
## EXISTING CONDITIONS BEFORE THE INCURSION
Touch the box you wish to select.

| CONTROL TOWER | TAXIING PLANE | |
|---|---|---|
| DEPARTING PLANE | AIRPORT CONDITIONS | SEE INCURSION AGAIN |

NEXT

AUDIO

| Condition | Go To | Programming Notes | Display Notes |
|---|---|---|---|
| Control | 301 | Users who have not answered | |
| Departing | 340 | questions 3A-3D (660,670,680, | |
| Taxiing | 320 | 690) correctly will see only | |
| Airport | 360 | this menu--unless they choose | |
| EXIT | 10 | to override it. | |
| Next Menu | 400 | | |
| Incursion | 200 | NEXT MENU option only appears | |
| | | if user has seen but not | |
| | | answered all four questions | |
| | | correctly.  USER AUTOMATICALLY | |
| | | GOES TO 400 AFTER ANSWERING | |
| | | ALL FOUR QUESTIONS CORRECTLY. | |

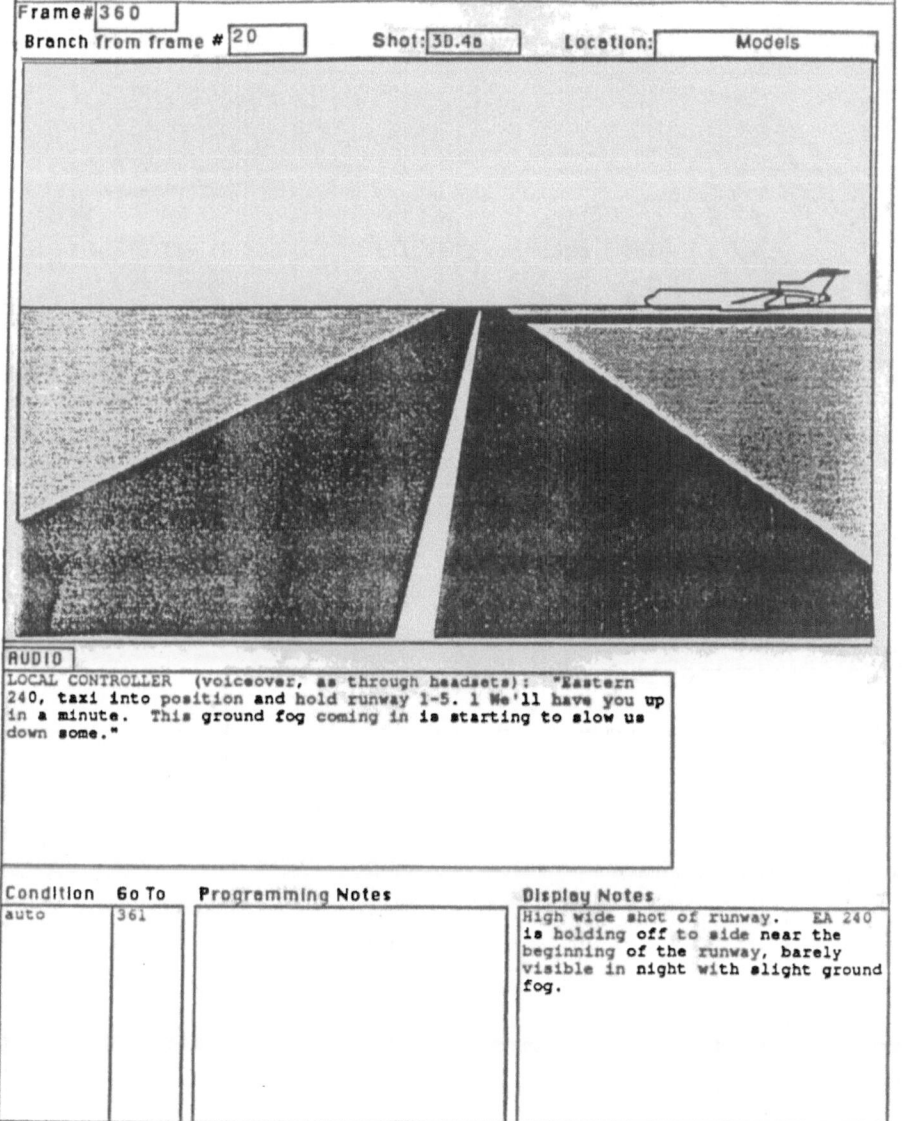

Frame# 360

Branch from frame # 20        Shot: 3D.4a        Location:        Models

AUDIO

LOCAL CONTROLLER (voiceover, as through headsets): "Eastern
240, taxi into position and hold runway 1-5. 1 We'll have you up
in a minute.  This ground fog coming in is starting to slow us
down some."

| Condition | Go To | Programming Notes | Display Notes |
|-----------|-------|-------------------|---------------|
| auto | 361 | | High wide shot of runway.   EA 240 is holding off to side near the beginning of the runway, barely visible in night with slight ground fog. |

Frame# 361
Branch from frame # 360        Shot: 30.2a        Location:    EA 240, exterior

AUDIO

COPILOT EA 240 (as through headsets; to pilot):   "Move into
position."

PILOT EA 240 (as through headsets):   "Roger that."

| Condition | Go To | Programming Notes | Display Notes |
|-----------|-------|-------------------|---------------|
| auto | 362 | | Two shot of pilot & copilot through windshield. |

Frame# 362
3ranch from frame # 361      Shot: 30.2b      Location: EA 240, interior

AUDIO
PILOT EA 240: "I don't know. Damn ground fog's always a
problem at this airport. Just three miles visibility. Hmmm."

| Condition | Go To | Programming Notes | Display Notes |
|-----------|-------|-------------------|---------------|
| auto      | 363   |                   | OTS           |

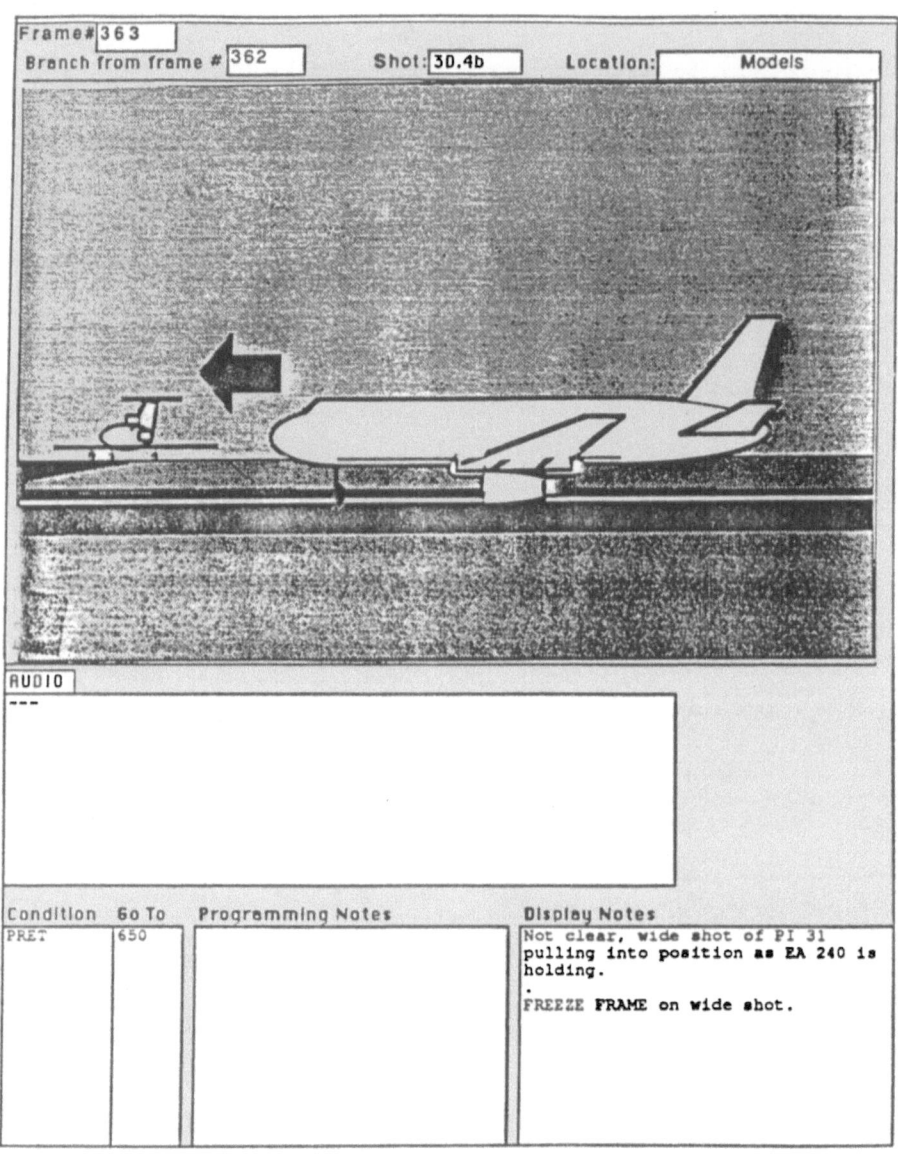

Frame# 363
Branch from frame # 362        Shot: 30.4b        Location:        Models

AUDIO
---

| Condition | Go To | Programming Notes | Display Notes |
|---|---|---|---|
| PRET | 650 | | Not clear, wide shot of PI 31 pulling into position as EA 240 is holding. · FREEZE FRAME on wide shot. |

Frame# 650

Branch from frame # 363                                     Question # 3D

X

Please select the best answer by touching the appropriate box.

A contributing factor to the Incursion occurred
In the general airport conditions.   Please Identify It.

a.   The ground fog made visibility difficult.

b.   Terminal buildings prevent the Controllers from
     having a clear view.

c.   The runways are so short that the  departing
     planes have to accelerate too quickly for the
     crews to pay attention.

| SEE AGAIN | EXIT | MENU | ← | → |

AUDIO

| Condition | Go To | Programming Notes | Display Notes |
|---|---|---|---|
| a-corr | 652 | Chosen answer box turns green when touched by learner. | |
| b | 654 | | |
| c | 656 | | |
| See Again | 360 | | |

Frame #'s  652,654,656

Branch from frame # 650                              Feedback to Question # 3D

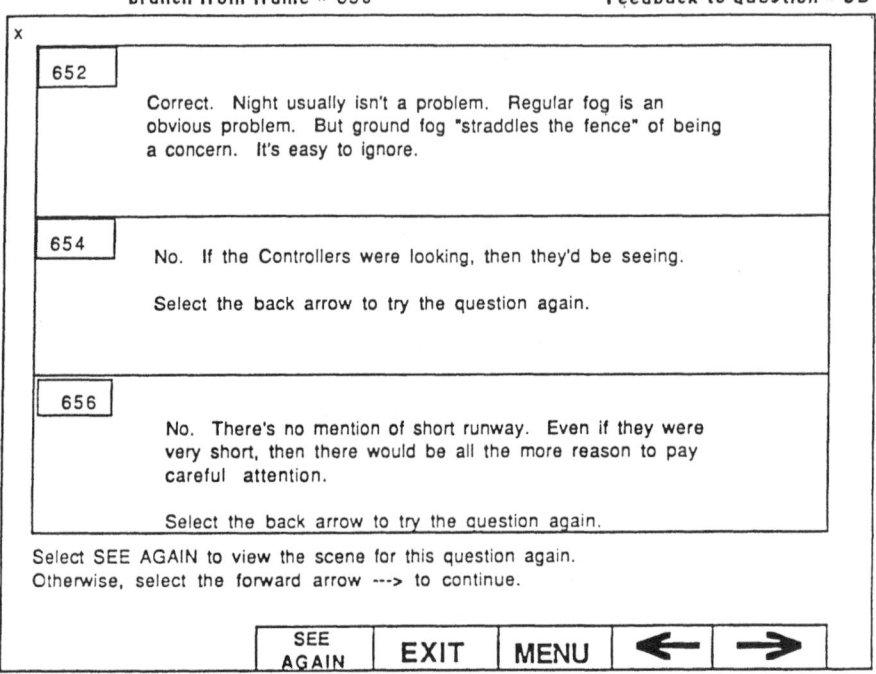

x

652

Correct.  Night usually isn't a problem.  Regular fog is an obvious problem.  But ground fog "straddles the fence" of being a concern.  It's easy to ignore.

654

No.  If the Controllers were looking, then they'd be seeing.

Select the back arrow to try the question again.

656

No.  There's no mention of short runway.  Even if they were very short, then there would be all the more reason to pay careful attention.

Select the back arrow to try the question again.

Select SEE AGAIN to view the scene for this question again.
Otherwise, select the forward arrow ---> to continue.

| SEE AGAIN | EXIT | MENU | ← | → |

AUDIO

| Condition | Go To | Programming Notes | Display Notes |
|---|---|---|---|
| See again | 360 | New User forwards to 20 (Existing conditions Menu) if not correctly answered all four existing conditions. | |
| <--- | 363 | | |
| ---> | 20 or | | |
| | 400 | Experienced User forwards to 400 (Modifications Menu) if correctly answered all four existing condition questions. | |
| | | If this question answered correctly, place check on MENU | |

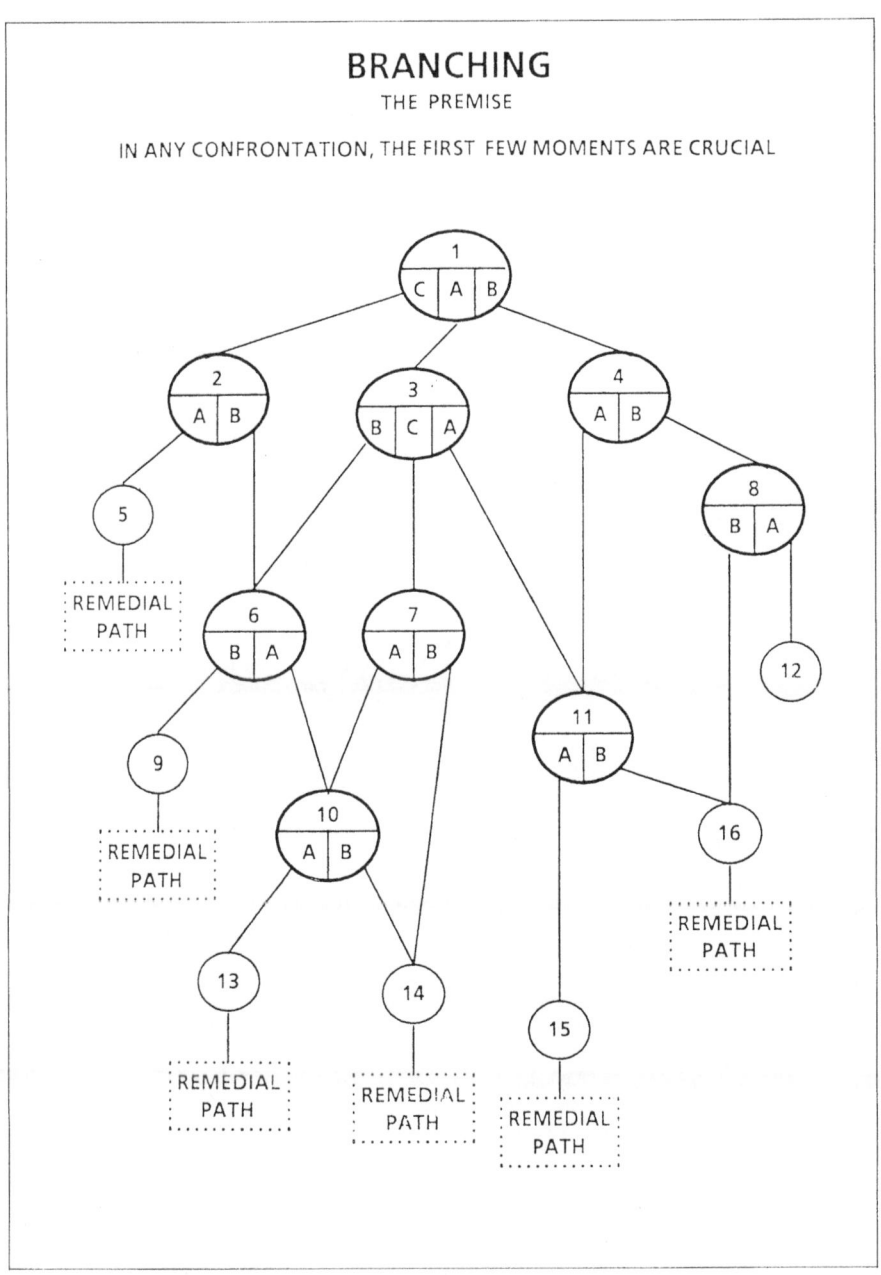

# BRANCHING

THE PREMISE

IN ANY CONFRONTATION, THE FIRST FEW MOMENTS ARE CRUCIAL

(Courtesy of Benjamin S. Walker and the U.S. Postal Management Academy.)

M A S T E R   S C R I P T

for an interactive videodisk program

THE PROBLEM WITH SHIRLEY

Produced for

POSTAL MANAGEMENT ACADEMY

by

Benjamin S. Walker
11317 Marcliff Road
Rockville, Md. 20852
(301) 493-5556

October 7, 1986

<u>**THE PREMISE**</u>

A black frame.  White titles FADE IN.
MUSIC:  An ominous drum beat, building tension.


PREMISE

In any confrontation,

the first few moments

are crucial.


FADE OUT AND GO DIRECTLY TO FRAME 1.

## FRAME 1

In a sequence of short shots, we see SHIRLEY throw some item to the Postal floor and stalk her way to the Supervisor's office. SUPER titles in sync with words:

NARR (VO):  A problem is headed your way:  Shirley Watson.  Age: 32.  Divorced, two children.  Present position: mailhandler.  Length of service: 8 years. She works for you.  Her problem: she was passed over for promotion.  Now it's your problem.

AGE: 32
DIVORCED
CHILDREN: 2
POSITION: MAILHANDLER
SERVICE: 8 YEARS

MUSIC CONTINUES UNTIL SHIRLEY SPEAKS.

Then CUT to a shot of SHIRLEY bursting through the door, into an office, and right up to the desk.  She leans over the desk and talks directly to the camera (representing YOU, the Supervisor).

SHIRLEY (VERY ANGRY):  I just saw the list for new supervisors and my name's not on it.  Can you explain that to me?

FREEZE FRAME.  POP ON INSTRUCTION TITLE:

    PRESS SCREEN
    TO INDICATE
    YOUR RESPONSE

THEN POP ON RESPONSE TITLES:

        A       SUPPOSE YOU START OVER

        B       WHOA, SHIRLEY -- WHAT'S THE PROBLEM?

        C       WHO LET YOU IN HERE?

Branching

    1A   -   3

    1B   -   4

    1C   -   2

### FRAME 2

Shirley listens,
then answers.
She still leans
over desk, is still
very angry.

YOU (VO -- ALSO ANGRY):   Who let you
in here?

SHIRLEY:  My 8 years of good service let
me in here, that's who.   I want you to
tell me why I was passed over for
supervisor.

FREEZE FRAME.   POP ON RESPONSE TITLES:

    A    DON'T TELL ME WHAT TO DO

    B    IT'S A LONG STORY

Branching

    2A  -  5

    2B  -  6

4

<u>FRAME 3</u>

Shirley straightens          YOU:  Suppose you start over.  What's
up.  She is still
angry.                       the problem?


                             SHIRLEY:  The problem is that after

                             8 years of good service, you passed me

                             over for promotion to supervisor.  And I

                             want to know why.



FREEZE FRAME.  POP ON RESPONSE TITLES:

    A    SIT DOWN, SHIRLEY.  LET'S TALK ABOUT IT

    B    YOU DID IT TO YOURSELF

    C    YOU'RE NOT THE ONLY ONE PASSED OVER


<u>Branching</u>

   3A  -  11

   3B  -   6

   3C  -   7

5

<u>FRAME 4</u>

Shirley straightens          YOU:    Whoa, Shirley!   What's the problem?
up.  She is still
angry, but calmed            SHIRLEY:  I just told you -- after 8 years
down a little.
                             of good service you passed me over for

                             supervisor.  I want to know why.


FREEZE FRAME.  POP ON RESPONSE TITLES:

    A    SIT DOWN, SHIRLEY.  LET'S TALK

    B    SIT DOWN, SHIRLEY.  I'VE BEEN
         WANTING TO TALK TO YOU


<u>Branching</u>

    4A  -  11

    4B  -   8


6

FRAME 5

Shirley is extremely          YOU (ANGRY):   Don't tell me what to do.
angry.  After
delivering lines,             SHIRLEY:  I'll tell you what I'm going
she stomps out of
the room.                     do -- and you'll find out from the union.


FADE OUT/FADE IN

Narrator, in an office setting, talks to camera.

                              NARR:  You didn't handle this situation

                              well.  Let's review what happened when

                              Shirley came through your door and see

                              how you might have done better.


Branching

   1CR - 2AR - 5X

7

FRAME 6


Shirley straightens          YOU:  It's a long story.
up.  She is still
angry.                       SHIRLEY:  I've got time.  Suppose you

                             tell me why I'm not good enough to be

                             a supervisor?


FREEZE FRAME.   POP ON RESPONSE TITLES:

        A     I DON'T HAVE ALL THE FACTS

        B     SUPPOSE WE DO IT LATER


Branching

        6A  -  10

        6B  -   9


                              8

FRAME 7


Shirley straightens          YOU:  You're not only one that was
up.  She is still
angry.                       passed over.

                             SHIRLEY:  I don't care about other

                             people.  I want to know why I was

                             passed over.



FREEZE FRAME.  POP ON RESPONSE TITLES:

        A    I DON'T HAVE ALL THE FACTS

        B    SIT DOWN, SHIRLEY.  LET'S TALK



Branching

      7A  -  10

      7B  -  14



9

FRAME 8

Shirley sits down.          YOU:   Sit down, Shirley.  I've been
She is calmer.
                            wanting to talk to you.


                            SHIRLEY:   Then how come I have to barge

                            in here like this?   Just tell me why

                            I didn't get the promotion.



FREEZE FRAME.  POP ON RESPONSE TITLES:

        A    LET'S LOOK AT THE FACTS

        B    I KNOW YOU ARE ANGRY



Branching

        8A  -  12

        8B  -  16


10

<u>FRAME 9</u>

Shirley is still       YOU:  Suppose we do it later.
very angry.

                  SHIRLEY:  If we do it much later,

                  you'll be talking to my union steward.


FADE OUT/FADE IN


Narrator, in an office setting, talks to camera.

                  NARR:  You didn't handle this situation

                  very well.  Let's review what happened

                  when Shirley came through your door and

                  see how you might have done better.


<u>Branching</u>

   IF INITIAL PATH:      THEN GO TO REMEDIAL PATH:

   1C - 2B - 6B         1CR - 2BR - 6BR - 9X

   1A - 3B - 6B         1AR - 3BR - 6BR - 9X


11

<u>FRAME 10</u>

Shirley stands             YOU:    I don't have all the facts
there.  She is
still angry.               right now.

                           SHIRLEY:   Why not?  You sure had enough

                           facts to deny me my promotion.


FREEZE FRAME.  POP ON RESPONSE TITLES:

        A    I DIDN'T DENY YOU ANYTHING

        B    SIT DOWN, SHIRLEY.  LET'S TALK


    <u>Branching</u>

      10A  -  13

      10B  -  14

                              12

<u>FRAME 11</u>

Shirley sits down.          YOU:  Sit down, Shirley —— let's talk.
She is a little
calmer.                     SHIRLEY:  The only thing I want to  talk

                            about is why I didn't get promoted.


FREEZE FRAME.  POP ON RESPONSE TITLES:

        A    LET'S TALK ABOUT YOUR ATTITUDE

        B    I KNOW YOU ARE ANGRY


    <u>Branching</u>

       11A  -  15

       11B  -  16

13

<u>FRAME 12</u>

Shirley settles                    YOU:   Let's look at the facts.
back in chair.
She is calmed
down now.                          SHIRLEY:   That's all I ask for.

                                   You'll see that I should be promoted.


FADE OUT/FADE IN


Titles against a black background:


                    Congratulations!
                    You handled this
                    situation just right.


FADE OUT


        PROGRAM STOPS


14

FRAME 13

Shirley is still           YOU:  I didn't deny you <u>any</u>thing.
angry.
                           SHIRLEY:  You're not even trying to

                           see my side.  I don't know what I'm

                           supposed to do.


FADE OUT/FADE IN


Narrator, in an office setting, talks to camera:

                           NARR:  You didn't handle this situation

                           very well.  Let's review what happened

                           when Shirley came through your door and

                           see how you might have done better.



Branching

        IF  INITIAL PATH:          THEN GO TO REMEDIAL PATH:

        1A — 3B — 6A — 10A         1AR — 3BR — 6AR — 10AR — 13X

        1A — 3C — 7A — 10A         1AR — 3CR — 7AR — 10AR — 13X

        1C — 2B — 6A — 10A         1CR — 2BR — 6AR — 10AR — 13X




                              15

<u>FRAME 14</u>

Shirley sits down.              YOU:  Sit down, Shirley -- let's talk.
She is calmer.

                                SHIRLEY:  It's a little late for that.

                                You should have been talking to me

                                before you stopped my promotion.


FADE OUT/FADE IN


Narrator, in an office setting, talks to camera.

                                NARR:  You could have handled this

                                situation better.  Let's review what

                                happened when Shirley came through your

                                door and see what you might have done

                                differently.


<u>Branching</u>

        IF  INITIAL PATH:            THEN GO TO REMEDIAL PATH:

        1C - 2B - 6A - 10B          1CR - 2BR - 6AR - 10BR - 14X

        1A - 3B - 6A - 10B          1AR - 3BR - 6AR - 10BR - 14X

        1A - 3C - 7A - 10B          1AR - 3CR - 7AR - 10BR - 14X

        1A - 3C - 7B               1AR - 3CR - 7BR - 14X


                    16

<u>FRAME 15</u>

Shirley is calmed
down quite a bit.

YOU:   Let's talk about your attitude.

SHIRLEY:   Sure I'm mad -- you would be,
too, if you were denied your promotion.

FADE OUT/FADE IN

Narrator, in an office setting, talks to camera.

NARR:   You handled this situation pretty
well.   But let's review what happened
when Shirley came through your door and
see how you could do even better.

<u>Branching</u>

| IF INITIAL PATH: | THEN GO TO REMEDIAL PATH: |
|---|---|
| 1A - 3A - 11A | 1AR - 3AR - 11AR - 15X |
| 1B - 4A - 11A | 1BR - 4AR - 11AR - 15X |

17

FRAME 16

Shirley is                          YOU:  I know you are angry.
pretty well
calmed down.                        SHIRLEY:  Sure I'm angry.  I feel like

                                    I should have been promoted and I wasn't.

                                    Maybe you call tell me why.

FADE OUT/FADE IN

Narrator, in an office setting, talks to camera.

                                    NARR:  You handled this situation very

                                    well.  But let's review what happened

                                    when Shirley came through your door and

                                    see how you could do even better.

Branching

        IF INITIAL PATH:            THEN GO TO REMEDIAL PATH:
        1A - 3A - 11B               1AR - 3AR - 11BR - 16X
        1B - 4A - 11B               1BR - 4AR - 11BR - 16X
        1B - 4B - 8B                1BR - 4BR - 8BR - 16X

                            18

REMEDIAL FRAMES


1AR          Repeat FRAME 1, as Shirley says:  I just saw the list
             for new supervisors and my name's not on it.  Can you
             explain that to me?

             NARR (LS--full screen):  You should have tried to calm
                  Shirley down a little before asking about her
                  problem.  What you said was ...


1BR          Repeat FRAME 1, as Shirley says:  I just saw the list
             for new supervisors and my name's not on it.  Can you
             explain that to me?

             NARR (LS--full screen):  You had a good answer ...


1CR          Repeat FRAME 1, as Shirley says:  I just saw the list
             for new supervisors and my name's not on it.  Can you
             explain that to me?

             NARR (LS--full screen):  You should have calmed Shirley
                  down and asked her what the problem was.  But
                  you said ...

2AR          Repeat FRAME 2:   YOU:  Who let you in here?
                               SHIRL:  My 8 years .... supervisor.

             NARR (LS--full screen):  You might have said something
                  to keep the conversation going, but you answered ...

2BR          Repeat FRAME 2:   YOU:  Who let you in here?
                               SHIRL:  My 8 years ... supervisor.

             NARR (LS--full screen):  You kept the conversation
                  going by saying ...


19

6AR          Repeat FRAME 6:    YOU:  It's a long story.
                                SHIRL:  I've got ... supervisor?

             NARR (LS--full screen):  You showed some involvement
                in Shirley's problem when you said ...


6BR          Repeat FRAME 6:    YOU:  It's a long story.
                                SHIRL:  I've got ... supervisor?

             NARR (LS--full screen):  You missed a chance to show
                some involvement in Shirley's problem.  Instead,
                you told her ...


7AR          Repeat FRAME 7:    YOU:  You're not ... passed over.
                                SHIRL:  I don't ... passed over.

             NARR (LS--full screen):  At this point, you could
                have invited Shirley to sit down and talk about
                her problem.  But you said ...


7BR          Repeat FRAME 7:    YOU:  You're not ... passed over.
                                SHIRL:  I don't ... passed over.

             NARR (LS--full screen):  You were on the right track
                when you said ...


8BR          Repeat FRAME 8:    YOU:  Sit down ... talk to you.
                                SHIRL:  Then how ... the promotion.

             NARR (LS--full screen):   The best thing to say at
                this point is, "Shirley, let's look at the facts."
                But you did show understanding when you said ...


9X           Repeat FRAME 9:    YOU:  Suppose we do it later?
                                SHIRL:  If we ... union steward.

             NARR (LS--full screen):  She's right.  If only you
                had handled the situation better.  Would you
                like another try?  Here's your chance.

             Camera PANS to a neutral screen.  POP ON response titles.

             A  TRY AGAIN    (This branches to point in FRAME 1
                                where Shirley comes through office door)

             B  STOP FOR NOW  (This ends the program)


                         21

10AR        Repeat FRAME 10:    YOU:   I don't have ... right now.
                                 SHIRL:   Why not? ... my promotion.

                NARR (LS--full screen):   Now would have been a good
time to ask Shirley to sit down and talk.   What
you said, though, was ...

10BR        Repeat FRAME 10:    YOU:   I don't have ... right now.
                                 SHIRL:   Why not? ... my promotion.

                NARR (LS--full screen):   You hit just the right note
when you answered ...

11AR        Repeat FRAME 11:    YOU:   Sit down, Shirley.   Let's talk.
                                 SHIRL:   The only ... get promoted.

                NARR (LS--full screen):   You could show some
understanding here by recognizing Shirley's
anger and telling her so.   But what you
said was ...

13X         Repeat FRAME 13:    YOU:   I didn't ... anything.
                                 SHIRL:   You're not ... to do.

                NARR (LS--full screen):   The situation is up in the
air at this point and about to turn into an
argument.   Would you like to start over from
the beginning?   Here's your chance.

                Camera PANS to a neutral screen.   POP ON response titles.

                A     TRY AGAIN       (This branches to the point in FRAME 1
where Shirley comes through office door)

                B     STOP FOR NOW   (This ends the program)

14X   Repeat FRAME 14: YOU: Sit down ... let's talk.
           SHIRL: It's a ... my promotion.

       NARR (LS--full screen): This situation isn't hopeless,
       but you could have been much closer to a solution
       by this time. Would you like to try again?
       Here's your chance.

       Camera PANS to a neutral screen. POP ON response titles.

       A TRY AGAIN  (This branches to point in FRAME 1 where
               Shirley comes through office door)

       B STOP FOR NOW (This ends the program)

15X   Repeat FRAME 15: YOU: Let's talk ... attitude.
           SHIRL: Sure I'm ... your promotion.

       NARR (LS--full screen): You handled this situation
       pretty well, but you're still not getting to the
       heart of Shirley's problem. Would you like to
       try again? Here's your chance.

       Camera PANS to a neutral screen. POP ON response titles.

       A TRY AGAIN  (This branches to point in FRAME 1 where
               Shirley comes through office door)

       B STOP FOR NOW (This ends the program)

16X   Repeat FRAME 16: YOU: I know you're angry.
           SHIRL: Sure I'm ... tell me why.

       NARR (LS--full screen): You're handling this
       situation extremely well -- the only thing you
       didn't do was get to the actual facts of the
       case. Would you like another try? Here's your
       chance.

       Camera PANS to a neutral screen. POP ON response titles.

       A TRY AGAIN  (This branches to point in FRAME 1 where
               Shirley comes through office door)

       B STOP FOR NOW (This ends the program)

23

Thus, in frame 2, two possible responses are offered: A, Don't tell me what to do, and B, It's a long story. If the supervisor punches A, the program branches to frame 5; if B, to frame 6, and so on.

Note that a "Remedial Path" leads from frame 5. Note, too, that several frames (3, 6, and 7, for example) are repeated, but not necessarily with the same audio.

What does Ben Walker, as a thoroughly experienced writer of interactive, have to say about all this?

*Ben, you've been writing scripts since 1966—about 500 of them at last count. But interactive has moved into the scene in just the past few years. Even with all your experience behind you, how did you learn to write for interactive?*

*Walker:* Well, I was interested in it. The Postal Service had a project. I got the assignment and I learned on the job.

*How did you go about planning* Shirley?

*Walker:* I drew it on paper. In other words, I knew that I wanted to have a number of different responses—a number of different avenues that the action was to follow. The worst one would be that Shirley would stomp out of the office and report her supervisor to the union. The best response would be she'd just sit down and say, "Well, all I wanted is a chance to talk. I'm glad that you'll talk to me." That's the best outcome. But there's a lot of difference between those. I came up with four other possible outcomes. Perfect went one way and the worst went the other way, but in between you might start off good but then make some bad choice and hit midway.

*What about the remedial angle?*

*Walker:* Oh, yes, you have to think about remedial. The computer has a great memory. Because of that you can say, "Well, here's what you did. You went from here to here to here. But let's look at what you might have done better." Then you show them a principle that they didn't adhere to, or that didn't occur to them. You give them the principle at each point where they made a worse choice than they could have made. After that you let them go through it again. It's somewhat like a rat in a maze. If you could tell the rat after he came out, "Now, look, here's what you did wrong. After two rights you should take a left," or something like that, hopefully, next time he'd follow the correct path.

A nice thing that I like about this, a secondary benefit for this medium, interactive, is that not only does it teach, but later it becomes a resource. If any viewers wonder, "What was it they said about that?" someone can go back in there and, as you modularize it, it's accessible like a book with different chapters. Using the menus, they can go back a month later when they've forgotten what you've taught them, but they'll remember that there was a section on, say, mortgage banking or something like that and they can refresh their memory, so it becomes not only a teaching tool but a resource.

*Once the planning is out of the way, is there any particular problem to the execution, more than there would be in a linear script?*

Walker: Well, there's the programming on the disc, and that's something I'm really not into as a writer. I have all the bits and pieces in there that they should shoot and after that it's a matter for the production team. I put it down on paper, like a roadmap, and then turn it over to the production people.

*Did someone work with you on this? I've been getting a lot of stuff about the instructional designer.*

Walker: I go counter current on this. All the instructional delivery service magazines talk about the team. There's the instructional designer, there are the research people, the writers, there are programmers, there are subject matter experts.

*But you—?*

Walker: I just did it myself. In other words I figured out what was the best way I thought to teach—always teaching but not boring somebody. I did it the way I thought was best.

*Is it unusual for the writer to do it this way? Is it common?*

Walker: I'll tell you, Dwight, I've read some books by people supposed to be experts on instructional design. But to me they're just using difficult terms to state obvious simple facts. You know how someone will over-construct and overstate to make something seem important. It just seems that many of these writers on instructional design follow that route. Maybe I'm just lazy, not willing to team up with others, but every time I get in fights with clients it's over this. Sometimes you're better off doing it yourself.

## COMPUTER-BASED TRAINING

Computer-based training (CBT), also termed *computer-aided learning (CAL)* and *computer-aided instruction (CAI)*, is another branch of interactive, but one that uses a computer in the actual instructional process. Our example is from another of Ben Walker's scripts. This one, *Oral Lesions Associated with HIV Infection*, was for the Veterans Administration.

Interactive, whether disc- or computer-based, has certain restrictions. The space on a disc can be used for seconds of audio, frames of video, or a combination of both. Different systems have different limitations as to amounts of recording space for sound, audio, and instructions. The complexity of the choices afforded also requires different technical considerations.

On this subject Ben Walker offers the best advice: "Don't get too tied up with equipment." As a writer, you need to be familiar with the capabilities and limitations of the system you're using for a specific project. But don't limit your thinking. Technology is advancing so quickly that the impossible today may be commonplace next week.

Always, however, the principles of good AV writing remain the same. If you've learned the guidelines in this book, you're on safe ground.

```
                    ORAL LESIONS ASSOCIATED
                      WITH HIV INFECTION


    TEXT AND SOUND                                    PICTURE

   (GRAPHIC MODE)                          Quarter-size picture
                                           of oral candidiasis
   Optional: Intro tune

   (Main title):

   ORAL LESIONS
   ASSOCIATED WITH
   HIV INFECTION

    (replace with):

           by

   Ralph W. Correll
      Director,
   Western Dental
   Education Center

        and

   William B. Wescott
   Chief, Dental Service,
   VAMC, San Francisco

   ------------
   (TEXT MODE)

   SOUND:  We'd like to know your name.

   Please enter your first name:

   Hello, $NS, welcome to an
   interactive tutorial on the
   oral manifestations of HIV
   infection.

   -------------
```

(Courtesy of Benjamin S. Walker and the Veteran's Administration Medical Center.)

This tutorial does NOT address the
complex medical considerations of
AIDS.  It is intended to make all
health care providers -- especially
those concerned with the oral health
of patients -- aware of the changes
that occur in oral and perioral
tissues as a result of infection by
the Human Immunodeficiency Virus (HIV).

Clinicians should understand the
importance of recognizing precursor
lesions that herald the onset of AIDS.

------------------

OBJECTIVES

After completing this tutorial
you should be able to:

SOUND: Announce items with "pings"

- List the oral manifestations
  of HIV infection and identify
  those that are precursors of AIDS

- Describe the clinical presentation
  of the oral manifestation of HIV
  infection

- Describe procedures for obtaining
  a definitive diagnosis of the oral
  lesions that develop as a
  consequence of HIV infection

- Explain treatment and management
  considerations for the oral
  manifestations of HIV infection

--------------
  MAIN MENU

A  Program
B  Review Menu
C  Quiz

X  Quit Program

-----------

```
                    REVIEW MENU

         A  Oral Infectious Diseases
         B     Oral Candidiasis
         C        Pseudomembranous
         D        Erythematous
         E        Chronic Hyperplastic
         F        Diagnosis
         G        Treatment
         H     Other Oral Infections
         I  Oral Hairy Leukoplakia
         J  Oral Neoplasms
         K     Kaposi's Sarcoma
         L     Squamous Cell Carcinoma
         M     Lymphoma
         O  Periodontal Manifestations of HIV
         P  Other Oral Diseases Assc. with HIV
         Q  Summary
         R  Quiz

         S  MAIN MENU
         X  Quit Program

         --------------------


                   RULES OF THE ROAD

         Whenever you see this information
         bar, (SHOW BAR AT BOTTOM OF SCREEN)
         the computer is waiting for your input.

            ESC - returns you to the Main Menu,
                  where you can quit the program

            F1  - returns you to the start of the
                  section you are presently in

            SPACE - advances you to the next frame

            ---------------
```

3

-- INTRODUCTION -------------

Evidence shows that specific oral
diseases are developing as a
consequence of HIV infection.

Oral lesions are common among AIDS
patients, probably because of the
impaired immune system and the
predilection for opportunistic
infections in these patients.

These maladies, which include
infectious diseases and neoplasms,
become manifest at differing times
during the infection.  Some arrive
early, prior to the clinical
manifestation of HIV infection.

Several of these oral diseases are
significant in that they foreshadow
the development of AIDS.


-- ORAL INFECTIOUS DISEASES -------

Fungal, viral, and bacterial infections
are associated with HIV infection.

Most are felt to develop as a consequence
of immune system suppression secondary to
HIV infection.

Oral candidiasis is probably the most
important of these infectious diseases
because it is considered a precursor
lesion to clinical symptoms of HIV
infection and, subsequently, AIDS.

4

```
-- ORAL CANDIDIASIS ------------

Candida organisms are a common fungal
inhabitant of the oral cavity,
GI tract, and vagina of many normal
persons, causing no apparent disease.

There must be penetration into the
tissue to cause infection, and this
occurs only under certain circumstances.

An increase in the incidence of
candidiasis has resulted from more
use of:

  - Antibiotics, which destroy the
    inhibitory bacterial flora

  - Immunosuppressive agents

  - Cytoxic drugs

  - Radiation therapy

Oral candidiasis is more frequently
seen in patients receiving chemotherapy
for leukemia, lymphoma, or other neoplasms.

Infants, debilitated patients - especially
the elderly, and patients that are
immunosuppressed are particularly
susceptible to candida infection.

Candidiasis is probably the MOST PREVALENT
opportunistic infection possible.

Therefore, $N$, it's no surprise to learn
that candidiasis is the MOST COMMON oral
infection reported in patients with
AIDS-related complex (ARC) or AIDS itself.

This common oral fungal disease is most
frequently caused by the organism Candida
Albicans.
```

```
-- ORAL CANDIDIASIS --------------------

Esophageal candidiasis has been included
in the WHO/CDC case definition for AIDS...

but oral candidiasis is NOT included,
EVEN THOUGH it is present in approximately
three-quarters of all patients with ARC or
AIDS.

In a study by Phelan and associates of
103 patients with AIDS, oral candidiasis
was almost universal.

In ARC and AIDS patients, oral candidiasis
frequently arises early, before the symptoms
of HIV infection become manifest.

Therefore, oral candidiasis is considered
to be predictive of HIV infection in risk
patients.

Three basic types of oral candidiasis have
been associated with HIV infection:

    - Pseudomembranous
    - Erythematous
    - Hyperplastic

Regardless of type, oral candidiasis related
to HIV infection generally persists for a
prolonged period and should be considered a
chronic infection.

-- QUESTION ---------------

Oral candidiasis is not included in
the WHO/CDC case definition of AIDS.

    A. True
    B. False

(Answer: A)

SOUND and TEXT:   Respond to answer.
```

6

```
(GRAPHIC MODE)

PSEUDOMEMBRANOUS CANDIDIASIS

SOUND: This is Pseudomembranous          Slide 1 - Title:
       candidiasis.
                                            Pseudomembranous
                                            candidiasis
Pseudomembranous candidiasis is
the type of Candidia infection
MOST FREQUENTLY found in patients
infected by HIV.

Also called "Thrush," it may
persist for months in AIDS
patients.

Although it commonly occurs on the
buccal mucosa, palate, and the
dorsal surface of the tongue, the
disease may involve ANY area of
the oral mucosa.
                                         DISSOLVE TO:

Clinically, it appears                   Slide 2 - Title:
as white-to-yellow curdlike
plaques on normal or reddish               Pseudomembranous
mucosa.                                    candidiasis with
                                           angular cheilitis
When the plaques are removed,
a hemorrhagic mucosal bed is
revealed.                                  MARK LOCATION OF
                                           ANGULAR CHEILITIS
SOUND:  Note the presence of
angular cheilitis.

Note the presence of angular
cheilitis.

(TEXT MODE)

-- QUESTION -------------

The white-to-yellow, curdlike
plaques of pseudomembranous
candidiasis can be stripped
from the mucosa.

     A. True
     B. False

(Answer: A)

SOUND and TEXT:  Respond to answer.
```

(GRAPHIC MODE)

ERYTHEMATOUS CANDIDIASIS

SOUND:  This is Erythematous
Candidiasis.

Slide 3 - Title:

    Erythematous
    candidiasis

The atrophic form of oral
candidiasis is referred to as
"Erythematous Candidiasis"
because of its red clinical
appearance.

It is usually multifocal,
commonly arising on the dorsum
of the tongue, palate, and
buccal mucosa.

When the tongue is involved,
the red lesions are frequently
located near the midline and
there is significant atrophy
of filiform papillae.

DISSOLVE TO:

Erythematous candidiasis of
the gingiva is UNCOMMON --
except in patients with
HIV infection.

Slide 4 - Title:

    Erythematous
    candidiasis of
    gingiva

CHRONIC HYPERPLASTIC CANDIDIASIS

SOUND:  And this is chronic
        hyperplastic candidiasis.

Slide 5 - Title:

    Chronic
    hyperplastic
    candidiasis

Oral lesions of chronic hyperplastic
candidiasis are uncommon in patients
seropositive for HIV.

When they do occur, they are most
frequently located bilaterally on
the buccal mucosa, especially in
the retrocommissural region.

The lesions generally consist of
firm, persistent red and white areas
that CANNOT easily be stripped from
the underlying mucosa.

8

*Ben, you wrote* Shirley, *which is an interactive disc. But you also wrote* HIV, *which is computer-based training. Is there any difference in the writing?*

Walker: *HIV* is for presentation on a computer, rather than on a video screen. Plus, the computer that presented *HIV* could talk. I wrote the things that it would say. They were recorded and put into computer voice files that could be called up by the program as needed. This particular script was heavy into pictures—pictures of AIDS lesions in the mouth. When we wanted to get as much picture on the screen as possible, I condensed the text and didn't say as much. I let the picture try to say as much as it could, rather than the words. The things I've done before, like for Met Life, for example, it's like all words except for a couple of Snoopy pictures to hold the audience. But in the AIDS thing you're showing pictures of AIDS lesions. You want to keep pictures up there all the time and use the sound to punch up important points and say words that the dentist might not have heard too many times. Like I always thought it was "KaPOsi" syndrome, when actually it's "KAposi" syndrome. Having sound, it touches another sense.

*Is there anything else to keep in mind?*

Walker: There's one thing you're constantly aware of, or at least that I was aware of, and that's not boring whoever's using the program. You're always aware of having something happen, something for them to do. I think I've coined a phrase for CBT shows—I call it *desktop theater.* I look at the computer screen and try to make a little theater in there—always something interesting happening.

*How complex should you get?*

Walker: Well, I know that there are great programs out there. For the individual freelancer, though, most of the big programs like ones that IBM does, for instance the one called *Reading to Write,* or the big ones for the Air Force, those are done by big teams. They take a year to do. It's hard for an individual to go out and get a complete CBT job to do. It's hard to know which authoring system or language to use, it's hard to get started. It's the old thing. The way I got into scriptwriting in the first place, I was interested in doing it and so I went out and started hanging around people who were doing it and I learned from them. So if you're interested in writing interactive, the computer-based training scripts, seek out the company or team or group that's doing it. With a little bit of preparation, a little bit of writing or instructional background, you might get on the team. If you go out and try and do it by yourself it's very difficult.

*It's changing fast, too.*

Walker: It's a field in flux, so far as I'm concerned. There's all kinds of CDI [compact disc interactive] and CD-ROM [compact disc, read-only memory] and that kind of thing coming along, but you have to think about what people have on their desks, and a lot of people don't have all those fancy things. I mean, they don't have a CD-ROM player that they can interact with. The newest thing now is CDI. The companies are building those little white boxes like Nintendo.

*Does that mean that a lot of individual scriptwriters are going to lose out?*

*Walker:* Not necessarily, but I'm not sure which things are going to go. I think the simple thing that plays on an IBM compatible or maybe the MacIntosh are the ones that are going to last. I'd try to get involved with programs that aim at the least common denominator rather than the fancy interactive things. The bottom line is, don't get tied up with the mechanics and the hardware and stuff like that. Once you have your principles of telling somebody something in an interesting and memorable way, really teaching it to him and then testing him to make sure he knows it, once you can do that, it doesn't matter what machines you've got, you can do it.

In conclusion, it should be noted that our reason for citing these three samples is because not all interactive procedures or scripts are the same. Each shop tends to have its own rules, its own foibles and caprices. It's a point to remember any time you start a job.

## GAMES TO LEARN BY

A related subject is that of the role of computer-based programs in creating game playing and simulation as teaching tools.

Why games and simulation? Because, as one authority points out, a game tends to hold interest much longer than does the average lecture.

A game is a process of manipulating things according to arbitrary rules lacking any function in real life. When the rules relate to real life, as in spelling in Scrabble or learning to recognize multiples of five in dominoes, then the game becomes a learning experience. Thus Atari's "Pole Position™ II" is a computer simulation for amusement only. But apply the same principles to a driver's education kiosk and you have a computer simulation used to educate. The enjoyment, the challenge, the stimulation can all be equal. But by definition games are useless enjoyment, whereas an educational game is supposed to serve a purpose.

A simulation, in turn, is an imitation of reality designed to teach participants how to act when faced with the real-life situation. A good example is the Link Trainer, which teaches student pilots how to fly without endangering them as reality would.

Teaching games, even if not so labeled, have always been with us. Hopscotch and jacks taught counting. Jigsaw puzzles taught shapes and color recognition. Songs such as "Old Macdonald" taught animal names and sounds.

The computer's virtue in this area is its speed. It makes possible complicated games with sound and movement. One of the results of this is to make games a popular teaching tool. Karen Tomlinson of AT&T says, for example, "Games are a good way of teaching almost anything. We use audiographic teletraining to teach customer service reps how to deal with clients on the phone. It's almost perfect simulation and the instructors can deliver the course from their living room with a computer, a modem, and software. It saves 49 weeks on the road. We also use a computer game called 'Corporate Challenge' to train our executives in AT&T protocol."

Unfortunately, few games and simulations are scripted. As Will Wright, developer of "Sim City," observes, "I do most of my planning in my head and what little I have on paper is almost unreadable. Actually the computer code is far more readable than what I tend to scribble on paper. . . . Basically, when I developed "Sim City" I just started playing around with the computer, creating a city. It was sort of a feedback process between me and the computer. I'd add a little thing to it and flow with the concepts when it came out of me—you know, writing a little more and more elaborate computer program. That eventually formed a direction in my head. I thought I could simulate a city with this. And I started reading up on cities and it sort of just pulled me in that direction."

The computer, in a word, is pretty much the master in games and simulation. The programmer is the creator, and the scriptwriter functions primarily as content specialist or instructional designer rather than as writer.

## TELECONFERENCING

You have a group of salespeople in New York, another in Chicago, a third in Los Angeles, a fourth in New Orleans, and a fifth in Toronto. They need periodic pep talks and updates on new products and modifications.

Or, foreign language is a required subject for high school students all over the state. They'll get more out of the course, school officials agree, if they can study a wide variety of languages with native speakers for instructors. But individual schools cannot afford to hire native-speaker teachers for a variety of languages.

Or, emergency room physicians all over the country need to watch a hands-on demonstration of a new technique for treating burn trauma. They will probably have questions to put to the people who have developed the new procedure.

How do you most efficiently handle these problems? The answer today often lies in what's known as *teleconferencing:* an audiovisual approach that links the viewer with participants across the country or even, increasingly, around the world.

What's teleconferencing? Define it as a meeting conducted via electronic communications, which enables participants in different locations to see and hear the same presentations and to respond to each other.

Most teleconferencing is made possible by global satellite television transmission. To oversimplify, satellites in space in a fixed position over the earth repeat back signals transmitted from the ground. These signals are received by "downlink" satellite dishes, which carry the program to local points as desired.

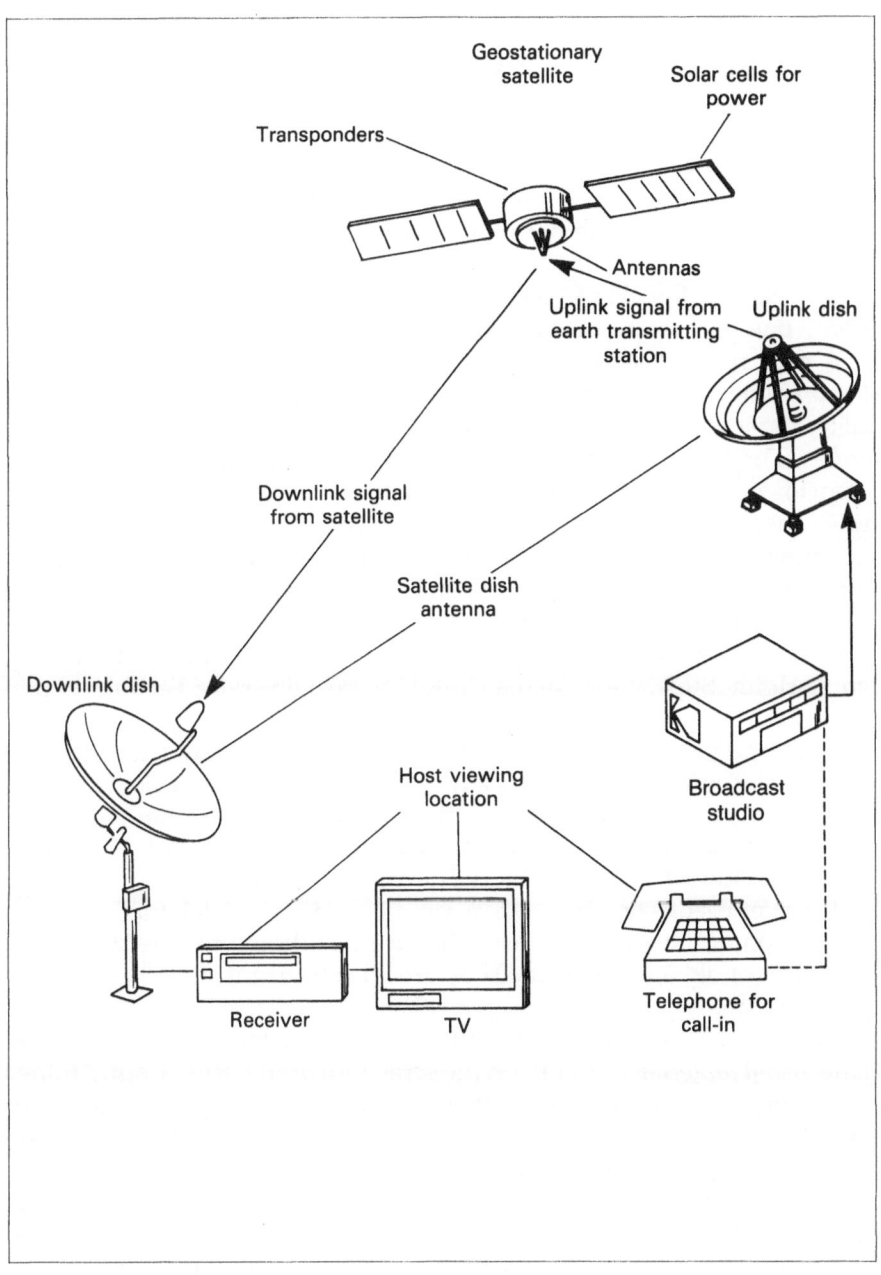

Teleconferencing comes in a variety of forms, each requiring different types of preparation. The simplest is audio teletraining. Both the U.S. Postal Service and the Veterans Administration use this for a variety of training purposes.

Here's the way it works. A training program, perhaps a linear video or a slide show or such, is prepared and shipped to local training sites. Students are brought together in a teleconferencing training room. There, a conference call connecting the training locations with the central training facility is set up. Speaker phones and push-and-talk microphones are used so only one participant can talk at a time.

A person at the local site is designated as the teletraining coordinator. At a predetermined time, everyone is in place. From the central training facility, the instructor conducting the lesson communicates over the speaker phones. The instructor tells the students what to look at. For example, the instructor might direct the local coordinator to play a videotape or a slide show. Then students are invited to ask questions.

Students hear the instructor and can talk with the instructor, but the instructor is not in the room. The Postal Service provides the students with a picture of the instructor, and the instructor has a picture of each student. "We find that helps them to relate to each other as individuals," says Dennis Holm, Supervisor, Media Unit, U.S. Postal Service Technical Training Center.

This type of arrangement is an example of a one-way teleconference or audio teleconference. It allows a wide variety of people to view the same material and to share their comments, questions, and responses.

"It's a poor man's way of doing teletraining, but it works," says the Veterans Administration's Richard Thorp.

Thorp envisions another step in remote-access video teletraining. "You have equipment in shipping cases that are distributed to sites where you are going to hold a conference. They are all self-diagnostic. They are very easy to set up.

"This allows the main-site instructor working a local console to control all the visual materials, to mark on the screen, to draft circles around things, to transmit real still pictures to these places. It gives the instructor real control over the conference and it's considerably cheaper than any other form of satellite teleconferencing."

The Hospital Satellite Network frequently uses one-way teleconferencing to combine previously prepared material with panel discussions or a question and answer session with experts. In the early stage of the teleconference, panel members may be asked previously planned questions. As part of their answer, a videotape, a slide show, graphs, or other previously prepared material may be used to illustrate the point the panel member wishes to make. Or the teleconference may begin with a video— say a tape of a delicate rhinoplasty operation performed by a plastic surgeon. After the presentation of the tape, the surgeon may be available to answer questions from the viewing audience.

"AT&T does a lot of teleconferencing," says Karen Tomlinson. "We have a lot of downlinks in our domestic organization. We can have a station, say, in Colorado, pull the signal off a satellite and distribute it across the network that's already in place. That's one of the advantages for AT&T. We just set up our 800 numbers for the interactivity. It's still a little awkward. You get 15,000 watching the same conference, some from as far away as Australia, for example, so when they call the 800 number maybe they get through and maybe they don't." But she's not discouraged. "I think the next great leap in technology is going to be totally interactive satellite transmission," she comments.

Two-way video teleconferencing is similar to what is seen on the evening news when speakers from different locations are shown on a screen to each other and to the viewing audience simultaneously. The participants in such discussions can see and hear each other and react to each other instantaneously. It's like a face-to-face discussion with participants separated geographically. Obviously, this is very expensive and very few organizations have the necessary financing or access to equipment to do this very often.

Once again, the amount of preparation for such a conference may vary greatly. As a general rule, the more significant and controversial the topic, the more tightly structured the preparation should be. Richard Thorp, Veterans Administration media executive, tells about potential problems:

> We have an expert panel from the VA. We show our canned presentations and then we open to questions from the viewers. With a controversial issue, the risk is that, under the pressure of the lights and the camera, panel members will find themselves setting policy. Whatever they say is the "word of God" as far as the viewer is concerned. That's a risk in live teleconferencing. It is almost impossible to undo a slip of the tongue.

> The way we handled this was, we recorded the callers, weeded out the kooks and crazies, and passed on the serious inquiries to the panel member most prepared to answer that type of question. Then we gave that panel member a few minutes to think about his or her answer and prepare. Only then did we broadcast the question as if it were arriving live.

The increase in teleconferencing in the last decade has been phenomenal. Any group that is widely scattered can find teleconferencing cost effective by saving travel expenses and time.

BASF Fibers of Williamsburg, Virginia, a major producer of fiber and yarn for commercial carpets, used to bring hundreds of carpet manufacturers, flooring contractors, specifiers, and end users to Williamsburg each year to tour the company. It was a costly and time-consuming way to inform clients about the company and its products. VideoStar Connections set up a special event broadcast that reached more than 550 of its most important customers via a 12-city interactive teleconference. "This gave us a vehicle to reach as many people in just two days as it would normally take us a year to reach," says Deborah Adams, national manager of contract merchandising for BASF Fibers.

American Express also uses VideoStar Connections services to establish an international network that includes downlinks across the United States and in Canada and England. This network is used primarily for employee communications and new product introductions.

Unisys Corporation has established Business Television, with a total of 90 downlinks in computer sales locations throughout the country. Customers can see a new product and then try it out themselves.

The United States Information Agency is a prime user of telecommunications. Through its Worldnet, it links over 150 U.S. embassies and consulates worldwide, and over 160 European and Middle Eastern cable, terrestrial, and closed circuit broadcasters. Its "Dialogue" public affairs program is described as a "live, unrehearsed, no-holds-barred international telepress conference," and it has been conducted in every inhabited continent. Journalists get tapes of the program and take sound bites from it for their television or radio stories, and print people can use it for newspaper stories.

"It's quite different from a commercial network," says Ann MacFarlane, a TV production specialist at Worldnet. "Our writers are well trained and professional, but they need to be trained in TV. We believe in involving your audience in what it's looking at. You don't just show a plane; you say 'in F-14s like this.' "

How is "Dialogue" handled? Here's how two Worldnet staffers see it: "We do one-hour talking head shows," explains associate producer Stacey Rose. "They're to serve a specific problem. Let's say people have a problem with growing coffee in Colombia. They need to talk to an expert American farmer who knows about growing coffee, so I include the most expert American and the embassy invites all the local farmers to the embassy. We do a lead-in video—a small presentation about the embassy that explains the purpose of this program. My scripts are minimal. I'm really only setting up the program.

Jim Randle, defense and foreign affairs correspondent, says,

We kind of background it, bring in the history. Bring the viewers up to what's happening and then we interview one or more guests.

It's shot live. We have to give our host the background and also a resumé about the expert; in case there's a problem or if he runs out of questions, he can use it. It's kind of like a press conference with an incredibly long mike cord. Or a seminar, a discussion. The beginning is scripted, but after that it's up to the guests and experts. The top of the program is carefully scripted, but the rest isn't scripted so much as it's outlined. Like most call-in shows, the show is controlled by the host.

The most common pattern is you have a significant guest on some pertinent topic and then there might be people at embassies all over South America involved. A guy in Paraguay may ask a question. Then a guy in Brazil asks

```
                    WORLDNET PROGRAM LOG          Page 2

Network/Program # EURONET 247  Worldnet # _440__  Date: THU 6/18/87

Sat. Time: Local: __8:50-10:00AM_ GMT: _1250-1400_____

Title: __100__  VIA SATELLITE/WORLDNET 1-LINE

Title: __103__  VIA SATELLITE/WASHINGTON

Title: __101__  HOST:  PAUL DUKE_____

Title: __102__  GUEST Dr FRANK OTTO,_____

                        EXECUTIVE DIRECTOR OF CALICO_____


***********************************************************************
                   1st     1st           Actual
Batting Order      Quest.  Block   Time   Time   Round Robin   Grid

 COPENHAGEN  _____ _____ 5 MINS _____ _____ _____

 ANKARA      _____ _____ 5 MINS _____ _____ _____

 BELGRADE    _____ _____ 5 MINS _____ _____ _____

 VIENNA      _____ _____ 5 MINS _____ _____ _____

 ATHENS      _____ _____ 5 MINS _____ _____ _____


***********************************************************************

_____  Back Time Close

_____  End Title Start    VTR# _____

_____  End Time

Title: _____  Goodnight                           (6849w)
```

(Courtesy of Worldnet Television, U.S. Information Agency.)

WORLDNET 440/EURONET 247
THURSDAY, JUNE 18, 1987
DOCUMENT 6849w
VTR ANIMATION

-3-

**ANNOUNCER**

Worldnet presents Dialogue, an
international televised exchange
of ideas. Now from our studios
in Washington, D.C., here is
your host, broadcast journalist,
Paul Duke.

STUDIO LIVE

**DUKE**

Hello and welcome to WORLDNET'S
DIALOGUE. For audiences in
Europe and participants in
Copenhagen, Ankara, Belgrade,
Vienna and Athens, we are
pleased to present a discussion
on new technologies in language
teaching.

Advanced technology is becoming
an integral part of language
teaching. Use of laser disks
and computers in the education
field is on the increase.
Language _is_ an essential human
activity, the main tool of
social and cultural
interaction. The need for
communication is great and there
are many organizations striving

-4-

STUDIO LIVE                    DUKE (cont.)

to meet this goal.  A leading

consortium in the effort to

design special software programs

for language learning is CALICO,

or Computer Assisted Language

Learning and Instruction

Consortium.  Founded in 1982,

CALICO now has 800 members

throughout the world, including

200 institutional members

ROLL VTR -                     and 50 corporate members.

TIME: 4:09

VTR                            NARRATOR

Technology has removed the limitations

of a book.  It continues to

evolve as a means of presenting

vast quantities of information

to quench man's thirst for

knowledge.  The ideal situation

for the teaching and application

of the language would be to

relocate the student body to the

target country.  Since it is

seldom feasible to transport

students all over the world, we

must artificially create an

effective version of the ideal

situation in our own back yard.

[Four pages of the script that give more background information are cut here. The next page completes the presentation of background information and moves into the conferencing part of the program.]

-9-

| VTR (cont.) | NARRATOR (cont.) |
| | implementation of other |
| | interactive video disc |
| | programs.  It works because it |
| | takes a student one step beyond |
| | the conventional classroom |
| | setting by transporting him into |
| | the lifestyle, language, and |
| | customs of a country. |
| | |
| STUDIO LIVE | DUKE |
| | To discuss this new technology, |
| | we are pleased to have as our |
| | guest Dr.Frank Otto, Executive |
| | Director of CALICO.  Dr. Otto, |
| | welcome to WORLDNET'S DIALOGUE. |
| | |
| | OTTO RESPONSE |
| | |
| | DUKE |
| | To begin today's discussion |
| | let's turn to our participants. |
| | Please remember to identify |
| | yourself and your organization. |
| | We will begin with Copenhagen. |
| | Go ahead please in Copenhagen. |
| | |
| SLIDES    GRIDS | FIRST DIALOGUE FROM COPENHAGEN. |
| _____    _____ | CONTENT: _____ |

-10-

STUDIO LIVE

DUKE

Thank you Copenhagen.  We will
now go to Ankara.  Go ahead
please in Ankara.

SLIDES     GRIDS

FIRST DIALOGUE FROM ANKARA

CONTENT: _____

DUKE

Thank you Ankara.  We will now
go to Belgrade.  Go ahead please
in Belgrade.

SLIDES     GRIDS

FIRST DIALOGUE FROM BELGRADE.

CONTENT: _____

DUKE

Thank you Belgrade.  We will now
go to Vienna.  Go ahead please
in Vienna.

SLIDES     GRIDS

FIRST DIALOGUE FROM VIENNA.

CONTENT: _____

a question. That kind of thing. The video is by satellite, but the questions are asked by telephone. Hosts must be prepared with questions in case the phone lines die. Sometimes people don't show up, so the host needs questions. The person in the control room on the coordination circuit will be saying "Brazil, get your people ready right now. The line to Paraguay just went down. Get on right now." Hosts must be prepared with questions in such cases. We write the questions.

Some of the programs become more elaborate, but that's the pattern. "Dialog" is presented unrehearsed. "Today we present a discussion on international economics or whatever, for an audience in this country and that country."

Actually, it's not as complicated as writing a news story. Because I don't have to tell the whole story.

You can start with the words or you can start with the pictures. Once I was assigned to cover a story about an attaché who had been killed by terrorists. They buried him at Arlington. It was a big affair, with lots of marching and a very high GBPS rating. That means Goose Bumps Per Square Inch. Very moving. I thought I shouldn't intrude, so I took the footage to an editor who was a former Marine and we selected the pictures and put together a picture story and I just narrated the slow parts and told only what was necessary—the attaché's name, why he was dead, and so on. My whole script ran only 10 or 15 minutes. You saw and heard real sound and heard taps at the end. I shut up and let the pictures talk.

In addition to news and public affairs materials, Worldnet uses many informational and entertainment programs. These are scripted like any video. *Tropical Rain Forests—A Disappearing Treasure* offers an example. This series on the environmental destruction of tropical rain forests was first aired in 1988–89 and was scripted by Jenifer Litwin.

Jenifer's opening step in preparing the Tropical Rain Forests series was gathering notes and ideas from periodicals.

Next came a preliminary audio script.

The final script used during the editing session is in the familiar two-column format, with video on the left, audio on the right.

"There exists a world where the air is thick mist, where tree frogs sing and where          ."
-covering seven percent of the world's land.
-INHABITANTS Home to an estimated 30 million species of plants and animals, many of which still remain unidentified.
Hundreds of thousands of indigenous people live in the tropical forests of the world, as hunter-gatherers, farmers, fishing people and collectors of forest products for sale or trade.  Tribal people have discovered many of the forest's secrets, and how to harvest its riches without abusing the privilege.
-RAINFALL Tropical forests are located wherever it is hot and humid year-round:(graphic of World with highlighted rainforest areas)Latin America, Africa, Asia, Australlia, etc..
-----Average annual rainfall is 60 inches, some receiving up to 400 inches.
TEMPERATURE The average annual temperature of the world's forests is 75 degrees Fahrenheit.

WHY ARE TROPICAL FORESTS IMPORTANT TO ALL OF US?
Many things we take for granted:
coffee, fruits (bananas,grapefruit,lemon,lime,orange,pineapple),peanut snacks, rubber used for many products like children's balls, anesthetics,nuts, spices.
Today, tropical plants form the chemical basis for many medicines:
pharmaceuticals such as Annatto(red dye) and curare(muscle relaxant for surgery)(many plants are valuable sources of food and pharmaceuticals).
One-quarter of the prescription drugs used in the U.S.A. # of major medicines in use today were isolated from plants.

GLOBAL ROLE OF RAIN FORESTS POINT TO SUCH VITAL CONTRIBUTIONS TO HUMANITY AS THESE:  1-source of countless thousands of useful plants and
animals.(one-quarter of the prescription drugs used in the U.S., including potent cancer and cardiac medications.)derived from or inspired by rain forest plants.  2-the lush vegetation of rain forests also plays an important role in modulating the earth's temperatures.
WHAT ARE THE CAUSES OF THE RAINFORESTS DISAPPEARING?
-Destruction continues at a rate of 50 to 100 acres a minute (largely for the benefit of the developed nations), a pace that could wipe out what's left by the end of the century; 50 MILLION ACRES OF TROPICAL FOREST ANNUALLY IS CLEARED (inflicting severe economic, social and environmental costs on 56 developing countries).
LOGGING
BUILDING OF NEW DAMS Large-scale hydroelectric dams.
CONVERSIONS OF FORESTS TO FARMLAND  and the forest is not being given enough time to grow back.
RUNAWAY POPULATIONS
SPREAD OF AGRICULTURE
DEFORESTATION
CATTLE RANCHING

Consequences: mass species extinction,further impoverishment of poor populations, permanent degradation of ecosystems, and the destruction of the world's food base.

Farming, logging, cattle ranching, and massive development projects (such as roads and hydroelectric dams) are the primary direct causes.  Indirect causes include worldwide demand for the forest's products, as well as the growing human population.

(Courtesy of Worldnet Television, U.S. Information Agency.)

HOW CAN FORESTS BE SAVED?
1-public concern and awareness.
2-reserves
3-sustained agriculture
4-new harvesting methods
5-restoring degraded land


RAIN FOREST EXHIBIT (3 MAJOR THEMES)
1-Ecology of the rainforest, its beauty and diversity
2-People of the rainforest and their interactions
3-Destruction of the rainforest, causes, consequences and possible solutions.

Script #1

There exists a
world where the air
is thick mist....
Where flowers
define tropical
beauty....
And where the
waters run
peacefully under a
canopy of
hundred-foot trees.
This is the world
of the Tropical - *covering 7%...")*
Rain forest.  These
forests, covering
seven percent of
the earth's land
mass, grow wherever
it is hot and humid
year-round.  But
tropical rain
forests, like South
America's Amazon,
are fast becoming
extinct.

(Courtesy of Jenifer Litwin, producer, Worldnet Television, U.S. Information Agency.)

They are being cut
down to meet the
needs of widespread
business
developments and
expanding
populations.  The
destruction of the
world's rain
forests is
threatening the
existence of an
estimated 30
million plant and
animal species that
live there.
Every year, nearly
20 million acres of
tropical forest are
cleared for
cultivation or
farming.  An
additional 12
million are cut for
timber, paper pulp,
and other wood
products.  And up
to five million
acres are destroyed
by cattle-ranching.

At this rate,
forests that took
ages to develop
will be gone in the
next century,
taking with them
those products
which we have
learned to depend
on. Rubber, which
is taken from
rubber trees, is
used for products
like children's
balls and shoes.
Medicines, derived
from snake venoms
and tropical plants,
are used in
prescription drugs,
helping to fight
off cancer, heart
disease, and other
ailments.

Shots, displaying these rainforest products

Products such as these are shown here are part of a major exhibition being organized by the Smithsonian Institution and the World Wildlife Fund. Entitled Tropical Rainforests: A Disappearing Treasure, this exhibit will travel across the U.S.

| | |
|---|---|
| Amazon Tape 2 7:46-7:51 | There exists a world where the air is thick |
| Amazon 27:44-27:47 coral flowers/dissolve to Science W. 2:30-2:37 | mist....Where flowers define tropical beauty....And where |
| Amazon 17:52-:56 boat/ dissolve to Amazon 5:39- 5:43 trees | the waters run peacefully under a canopy of hundred-foot trees.  This is the |
| Science W. 5:43-5:47 sky | world of the Tropical Rain forest.  Covering |
| Amazon :16 secs-:20 secs plane flying over forest/ dissolve to Science W. 3:46 | seven percent of the earth's land mass, these forests grow wherever it is hot and humid |
| Amazon 27:10-27:15 over- head shot of Amazon river | year-round.  But tropical rain forests, like South America's Amazon, are fast becoming extinct. |

*Handwritten notes:*
ROLAND
Wed. Am. - 8:00 - 9:00
Audio Room A

[ 2864 K ]

Nat sound on Cas. 3

Inmain script - The program will focus on the problems facing the

(Courtesy of Jenifer Litwin, producer, Worldnet Television, U.S. Information Agency.)

| | |
|---|---|
| Amazon 15:30–15:37 man cutting down tree<br><br>or Amazon 25:20–25:25<br><br><br><br><br>Amazon 5:14–5:20 plants/ cut to Amazon 15:32 frog<br><br><br><br><br><br><br><br><br><br>J.Kempf tape:30:12–:38:26 or Science W. 4:24–4:39 man plowing land<br><br><br><br><br><br>J.Kempf tape 58:11/<u>cut to Amazon 22:52–23:00</u> | They are being cut down to meet the needs of widespread business developments and expanding populations.  The destruction of the world's rainforests is threatening the existence of an estimated 30 million plant and animal species that live there.  Every year, nearly 7.4 million hectares of tropical forest are cleared for cultivation or farming.  An additional 4.8 million are cut for timber, paper pulp, and other wood products. |

| | |
|---|---|
| J.Kempf tape 1:22-1:27<br>cattle grazing | And up to two<br>million hectares<br>are destroyed by<br>cattle-ranching.<br>At this rate, |
| Amazon 11:48-11:57 car<br>driving down road/or<br>12:24-12:35 pan of baron<br>empty land | forests that took<br>ages to develop<br>will be gone in the<br>next century,<br>taking with them<br>those products<br>which we have<br>learned to depend |
| Cas 4 2:42 wide shot<br>products/cut to 3:10 balls/<br>cut to cu tennis shoe | on.  Rubber, which<br>is taken from<br>rubber trees, is<br>used for products<br>like children's<br>balls and shoes. |
| Cas 4 4:46 cu bottles | And medicines,<br>derived from snake<br>venoms and tropical<br>plants, are used in<br>prescription drugs,<br>helping to fight<br>off cancer, heart<br>disease, and other<br>ailments. |

---14---

# Multimedia Shows

If you want to let your imagination soar, multimedia shows offer incredible possibilities. They involve what might be termed a mixed bag. That is, they make use of a variety of media—still photos, film, live action, puppets, rear-screen projections, and so on. They coordinate these media into a single presentation.

Multimedia presentations range from a slide show accompanied by a live narrator to Universal Studio's Earthquake ride to Disney World's Catastrophe Canyon.

You might even stretch things to include a touring Saudi Arabian exhibit that covers 20,000 square feet. It offers a Bedouin tent constructed on a mound of authentic Saudi sand; Saudi guides in traditional robes; video kiosks presenting continuous shows on falconry, Bedouin life, Arabian horses, and such; models of buildings and towns; interactive videodisc kiosks that respond to questions about oil exploration and production in Saudi Arabia; linear videos projected in theater-like settings; a simulated marketplace with stalls and artisans; a life-size replica of a city street with balconies, shops, and homes; and a laser light and video show. It's impressive and carries the audience through from the origins of the country to today.

At the Huntsville, Alabama, Space Camp the audience enters a "space ship." Sounds of rocket blasts reverberate through the room and the floor vibrates in a simulated takeoff as a screen shows shots of Earth retreating.

Similarly, horror movies have sent fake "spectral wraiths" out from the stage to touch pleasurably terrified audience faces. Seats have vibrated and have even been wired to shock tension-addicted patrons electrically.

Sights, sounds, smells, and sensations (heat, cold, movement) can all be part of a multimedia presentation. Indeed, sooner or later someone's going to figure out how to simulate taste for the participants in a multimedia show!

In a word, the sky's no limit. Even though you'll probably start on less ambitious projects than those just mentioned, you'll still be faced with the question: How do you script a multimedia project?

As yet, no standard format has evolved—which means that, as is so often the case in AV, you're on your own, and good luck.

(That isn't to say that you may not use standard formats for appropriate segments. If a unit to be produced in film or video is incorporated, for example, you'll simply put it in that medium's format.)

The only issue in multimedia scripting, actually, is intelligibility. However you decide to tackle the problem, just be sure the people to whom you're going to present your work can understand what you're talking about.

To that end, use any or all of the techniques offered in this book along with whatever innovations of your own strike you as potentially effective. Sometimes, you may decide that a particularly detailed production script may do the job; other times, a storyboard or perhaps even some sort of graph or flow chart may be better.

At no time, however, should you forget the possibility that your best bet may be a clever treatment that captures the essence of your project *without* production details. As David Hon, respected West Coast expert, has observed, "Don't hesitate waiting for the technology. Design what you need, and the technology will be there. Technology is always well in advance of any use that's been made of it."

So charge ahead to the limits of your creativity. Get your idea down in whatever form makes it clear. It's often simply too much to expect that your sponsor will be able to follow a complicated shot list. It may be better to limit yourself to something like "Simultaneously we see stills of what's going on in each of the sports activity areas: bowling, tennis, swimming, diving, riding, racquetball . . . perhaps with a full-screen comedy topper of a turtle race, ping-pong, or a skateboarder falling into the pool." The details you can save for the producer. Whatever your project, in the final analysis you and your imagination are in command. In multimedia, as everywhere else in AV, you start with a topic, an audience, and a purpose; work up an idea, a concept, and a key point; search out a logical approach and a means of presentation; incorporate it all in a proposal, a treatment, and a production script; then cross your fingers and hope that the result will be cheers, not jeers.

Following are a sample script and production plan for a simple, interactive multimedia presentation located at National Geographic's Explorers Hall in Washington, D.C. It consists of a simulated space flight with viewers seated before a spinning globe of the earth.

Periodically, the globe stops and linear videos appear on two screens located on either side of the globe. A spotlight marks the region of the globe being discussed. The system asks the audience questions about the region and the audience members individually select answers from the choices

NATIONAL GEOGRAPHIC SOCIETY EARTH STATION ONE - SHORT SHOW
11/02/89          Filename: ES28NATM

Running Time:     approx. 11:31

1. Ladies and gentlemen welcome
aboard. I'm your captain and today
you're going to see Earth in a new
way:
[7:00/02:37:00]

2A. A living planet ~~whose geologic~~
~~activity shapes and reshapes earth;~~
~~seas, whose oceans currents~~
~~transport heat and regulate~~
~~climates; and great weather systems~~
~~circulate life-sustaining moisture.~~
~~A world~~ where a multitude of
systems maintain an improbable
balance critical to life on earth.
[15:29/02:54:28]

[Alternate end for para. 2A:
A world where multiple systems
maintain an astonishing balance -
a balance critical to life on earth.

~~2B. [Crew #1]~~
~~Captain.- Remote sensor network is~~
~~ready for activation.~~
~~[3:00/02:59:06]~~

2C. [Captain]
Thank you ensign.  Bring main power
grids on line.
[03:12/03:03:09]

~~2D. [Crew #1]~~
~~Aye Sir.- Coming on.~~

~~Power grids on line and phased, sir.~~
~~ESOSS section standing by.~~
~~[10:00/03:14:06]~~

~~2E. [Crew #2]~~
~~ESOSS direct sir;~~
~~--Multi-looped array~~
~~--Trans-global configuration~~
~~--All systems nominal.~~
~~-SensorNet on standby.~~
[07:10/03:22:23]

Sample script pages from the beginning of the interactive multimedia presentation.
(Copyright 1990. National Geographic Society.)

NATIONAL GEOGRAPHIC SOCIETY EARTH STATION ONE - SHORT SHOW
11/02/89          Filename: ES28NATM

2F. [Captain]

- Very good ESOSS, activate
  SensorNet
  ... Now.
  [03:12/03:27:06]

*[PAUSE]*

2G. [Captain]
Navigation.
[01:00/03:33:20]

2H. [Crew #2]
Navigation, sir.
[01:03/03:35:08]

2i. [Captain]
Let's take 'er out for a spin.
[01:25/03:37:18]

2J. [Crew #2]
Yes sir!
[00:29/03:38:27]

*[PAUSE]*

3.    3. Even from 23,000 miles in space
      we don't need our telescopic
      cameras-on remote sensors to
      appreciate the beauty of the
      planet.  It is vibrant and alive--
      friendly and inviting.
      [11:08/03:57:00]

4     4. Features on earth's surface
      reveal much about it's workings and
      its ability to support life.
      However even the largest things
      aren't always obvious.  The secret
      is knowing where to look.
      [11:05/04:09:08]

NATIONAL GEOGRAPHIC SOCIETY EARTH STATION ONE - SHORT SHOW
11/02/89          Filename: ES28NATM

**5. The single largest land feature
on earth cannot be seen from space.**

[04:14:08]
~~Quessub~~  **Do you know why?**
~~John.~~   IS IT

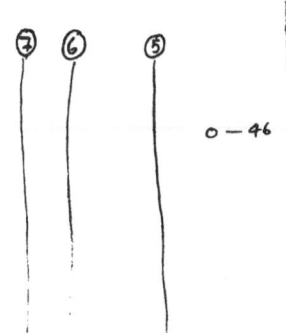

**. too small.
. hidden by clouds.
· covered by ice.**
OR **· underwater.**
[12:14/04:22:25]

**6.  Let me know what you think by
pressing one of the flashing
buttons in front of you.**
[03:13/04:27:05]

*[PAUSE]*

**7A. The world's largest land
feature is the Mid-Ocean Ridge, a
volcanic mountain chain beneath the
sea.**
[05:08/04:42:11]

o — 46

**7B. Our special lens allows us to
view it as it ambles around the
globe for 46,000 miles-- like the
seam on a baseball.  If we were to
place the Andes, the Rockies, the
Alps, and the Himalaya end to
end they still wouldn't be as
long as that ridge.**
[13:28/04:58:29]

o— 25

**8A.  In some places -- like Iceland
and the Azores -- this volcanic
ridge has broken the surface of the
water.**
[05:27/05:05:07]

**8B.  Here in the Atlantic, lava
flows from the ridge
- expanding the ocean floor.**

**N.G.S. Earth Station One Short Show - LOG OF ART TO BE CREATED**

Page 1

As of: 10/31/89

**Diagrams and Flat Maps**

| #of: Locale:  Size: Para /Sc | Description: (Format notes are in *Italic*) | Due | Stat |
|---|---|---|---|
| P.2G. C+D | Robinson map with probes and grid | | |
| P.3. A Locator | Robinson map with probes and grid: Comes off and on with locator display | | |
| P.4. C+D | Robinson map being scanned with features being filled.(either gets scanned and features follow in steps, or completely then map changes to map with features) | | |
| | Question and Answer Pad | | |
| | "Probe" Window (to house pictures) | | |
| P.7A. C | Physical Robinson map with Mid Ocean Ridge indicated. (Note: used in combination with 3D globe [transparent globe ?] See 3D models P 7B.) | | |
| P.11B A | Electric map of Pacific with highlighted trenches | | |
| P.12. A | Add plate lines to map P.11B (see above) | | |
| P.21. C | Earth horizon (over video shot) atmospheric detail showing no ultra-violet light passing through ozone layer | | |
| P.22. B | Robinson map with Ozone hole boundaries | | |
| P.23. A Locator | Flashes on probe display showing thunderstorm and stormmy weather | | |

Art logs corresponding to the script pages. There are four types of art employed in this presentation: diagrams and flat maps, 3-D models (animated), still photos, and short linear videos. Each item must be prepared and stored, generally on optical disk, for ready access by the system. The system orchestrates the presentation, drawing on the stored visual and audio components, according to the program established by the script. (Copyright 1990. National Geographic Society.)

**N.G.S. Earth Station One Short Show - LOG OF ART TO BE CREATED**

As of: 10/31/89                                                    Page 4

**3 D Models**

| #of: Locale:   Size:<br>Para /Sc | Description:<br>(Format notes are in *Italic*) | Due | Stat |
|---|---|---|---|
| A Locator | Locator - wireframe, use 24 polygons, outline continents and axis | | |
| Optional:<br>P. 3.A Replaces Locator or smaller ships position globe on B | Wireframe with probe super structure, outline continents, (or might be probe superstructure over rendered globe because two wireframes will be to confusing) | | |
| P.3.  B | Rendered globe with :<br>1) Glowing lines showing plates<br>2) Arrows indicating Oceans<br>3) Arrows indicating Jet streams and storm systems (Hurricanes) | | |
| P2. C+D | Wireframe "globe" with:<br>1) Glowing lines<br>2) Ocean system arrows<br>3) Arrows indicating Jet streams and storms | | |
| Screen B for Locator globe | Rendered globe with wireframe cube structure showing position of ship | | |
| P.7B. C | Globe either wireframed or with continents outlines or transparently rendered globe with Mid Ocean Ridge lines over Physical Robinson map (see P.7B Diagrams and Maps) | | |
| P.12  B | 3 D wireframed globe or transparently rendered globe showing major plates (Maybe in 3 or 4 position) | | |

displayed on the screen. The results of the voting are displayed and discussed.

This is an effective presentation, but is quite demanding for the writer(s) and production team.

# PART FOUR

# SUCCEEDING AS AN AV SCRIPTWRITER

---

$$15$$

# The Things You Sell

## GETTING STARTED

What's it like to be a novice scriptwriter? What are your problems? What steps do you take as you begin your career? To give you some sample answers, let's talk to Craig Guthery, a 27-year-old man from Oklahoma City getting his profession underway in the Washington, D.C., area. He had spent six years in the Navy as a nuclear power plant operator. Some of that time was in the training division. He had lived in a variety of places. By avocation he also was a rock musician, a vocalist. He had had three semesters of basic requirements at Central State University in Edmond, Oklahoma in 1979 and 1980. After he was discharged from the Navy he went to the University of Oklahoma in 1987 for a semester in film and radio. He had one course in film scriptwriting with Joye R. Swain, but he never did finish his degree. Then he went to Washington, D.C.

*How did you get started in audiovisual?*
*Guthery:* I read a newspaper ad that said a company was looking for someone with a degree in engineering who was familiar with film and video production, someone who could write for their audiovisual media. I applied. I didn't have the degree, but I had engineering experience, and in the scriptwriting class I'd scripted a fact film and a feature film. They liked what I'd done, the scripts I'd written for my class, so I got the job.

*How does your shop work?*
*Guthery:* I'm not sure that the job I have is a normal one. We don't have any production facilities. We have to borrow equipment. We don't even have a VCR, but for four years this small office that I'm working in has been producing videos. We're part of a big 2,400-employee corporation that works mostly on Defense Department jobs. For the most part, though, they have nothing to do with videos. What happened is that about four years ago this one small office of seven people had a contract to go out

and make some videos. They were going to subcontract out the whole thing, everything from the writing to the editing, and just get the final product back.

Well, they got the scripts back from the production house they'd hired. But this was an engineering video. It was kind of complicated. The writers from the production house had no engineering background. The script was very disjointed, impossible to use. I think maybe that was one of the reasons they hired me on—they thought that I could inject the creative part of the formula—good creative and yet informative technical video.

I'm one of the two people that work full time on AVs. We write the script, then subcontract. We subcontract for the cameraman and the post-production and all that, so we're producers as well as writers. We go down to the shoot and we don't hire a director, we have to do the directing ourselves. I've dealt with quite a few of the production houses and nobody really has a scriptwriter as a title on their staff.

*You do outside work too, don't you?*

*Guthery:* Yes. In fact, my first script was done freelance.

*How does that work?*

*Guthery:* Well, I met a writer. He's helped me a lot. He got a contract from the Post Office, and then I was able to do a couple of scripts for the Post Office through him.

*What was the first script you did for him?*

*Guthery:* A video script on laboratory services available in the VA hospital in D.C.

*What was the hardest thing about writing it?*

*Guthery:* I guess the hardest thing was knowing what things to ask the client. When I met with the Chief of Clinical Services at the laboratory I was as prepared as I thought I could be. I'd reread *Film Scriptwriting*. But when I got there I ran out of questions real quickly. I knew I had to have more information. Once I got started on the writing phase it was really difficult to translate this new information into a good, informative video script.

*What kind of thing was it that you lacked?*

*Guthery:* This video had to do with microbiology and hematology— terms I'd heard but that I hadn't known, if you know what I mean. While I was there I took a tour of the laboratory facilities where they'd be shooting, and from that point I learned most of what I put into the script. I learned it just from that initial meeting. And I did have a couple of follow-up meetings and phone calls to make sure that what I was writing was correct. I was afraid I hadn't done enough research, but I found out that sometimes even a superficial understanding can be translated into video.

And of course there's also the fact that when you lack information you ask the questions that somebody who's looking at the film might want to ask. Anyhow, I've been told that the script was a success.

*What other scripts have you done since you've been in Washington?*

*Guthery:* Well, on my job, my salaried job, I scripted a two-part troubleshooting training video for the Navy on high-pressure air compressors. Also, I've helped to rewrite three scripts by other people. All those drew a lot on my previous experience because I'd worked around high-pressure air compressors before. It was interesting to tie the scriptwriting and the air compressors together. I'm working on another assignment too. It's for the Defense Advanced Projects Agency. It's extremely interesting. They take some of the more futuristic ideas and see if they're feasible. A lot of it is engineering, but it kind of goes back to the laboratory script, because I don't know anything about what I'm writing about. Some of the programs are very complicated. Maybe it's just because I've done more scripts, but it seems to be coming a lot easier.

*Do you write at home or in an office of the company you're working for?*

*Guthery:* On the laboratory script, I actually did the writing at home. On the Navy scripts, I did the work at the office.

*How about atmosphere? Did people accept you as a writer?*

*Guthery:* At the laboratory everyone knew this was my first script. I think maybe they gave me a little more room to work with. Kim Luoma, a fellow who worked there, was very helpful. He was in the initial meeting with me. Asked some of the questions that I didn't, but I found that the Chief of Clinical was speaking more to him than to me.

*What about the Navy?*

*Guthery:* I already had one whole script behind me, so I figured that I was a pro now. I felt more confident. That helped a lot. I felt like this was something that I was familiar with, and I was going to write the best script that they'd ever seen.

*Were all your programs video?*

*Guthery:* Just straight linear video. In the training and industrial type of work it's almost exclusively video. I've found out that in my little corner of the corporate video field nobody even talks about film. I'm almost to the point of thinking that film is maybe only in Hollywood now. With the new formats, video is almost indiscernible from film anyway as far as the quality of resolution is concerned.

*How about interactive?*

*Guthery:* We had a shot at one of the contracts, but we didn't get it. Eventually we'll be getting into it, but not now.

*On your job, you're paid on salary. What about the others, the outside assignments? Are you paid by the script?*

*Guthery:* For my job, I'm on salary and I have benefits through the company. I bring home a paycheck every two weeks and it's a lot easier to budget. Then, for the other stuff, the individual scripts, I get paid by the assignment.

*What aspects of your work have been particularly useful?*

*Guthery:* Being on production was really helpful. I could see that something that looks like a great idea when you think of it in your mind is almost impossible to do in reality. A lot of shots that you think "Aw, these are really simple to do" can be very, very difficult. If scriptwriters could just go out and hang around the set—see what it takes to actually put a program on video or film, it would help an awful lot. If you could be there from the beginning and watch through the editing part of it and see what the work is like, you could learn what should have been left out in the first place and what probably should never have been written in the first place and ended up getting cut out in the editing.

*If you had it to do over, what would help you most? What would you like to have that you didn't have?*

*Guthery:* I'd like to have finished my degree—not so much for the diploma as for getting more information on film. I felt like a babe in the woods as I was going out into the field. I'd like to take another scriptwriting class. The class I took did a whole lot for preparing me; it was invaluable help. But if I took another, maybe I'd do better. I think that would be my first choice—if I could go back, I'd take more scriptwriting classes.

*How do you feel about the AV field in general now? Are you happy with it?*

*Guthery:* I'm very happy with it. I'm making money and I'm finally able to support myself writing video scripts.

And there you have a picture of the real world of AV, as seen by a beginning scriptwriter. It sums up a host of things you should know.

When you take your entry job, it is, as the saying goes, the first day of the rest of your life. More specifically, it's the beginning of your career. Whether that career will be long or short, happy or unhappy, will of course depend on many things. Talent, for example—whatever that is. Personality, ability to get along with people. Luck, good or bad.

## THE VITAL FACTORS

But beyond this, three other factors enter into your success—factors sometimes overlooked in the rush of things and preoccupation with the moment. Those three are insight, knowledge, and reliability.

### Insight

Some people assume that insight and knowledge are one and the same thing.

They're wrong. Insight is penetrating discernment. Knowledge is awareness of facts. Insight is also known as *judgment. Common sense* is another term that applies. It's the ability to appraise whatever is under consideration and relate it to reality—to think beyond the immediate to the big picture.

This characteristic is important to a scriptwriter, as witness a down-to-earth experience of a friend of mine. He was given a chance to script a video for the highway department. He did a fine job of it.

When he turned it in, though, the producer let him go.

Why? Because he'd specified shots from all four seasons. "Can you imagine?" said the producer. "It never even dawned on the idiot that that meant we not only couldn't meet the deadline, but that running out crews in winter with snow, spring with green, autumn with falling leaves, and so on would send the costs through the roof. I couldn't ever trust anyone with that little judgment."

## Knowledge

You already know about knowledge. It means having the facts of your job at your fingertips, right?

Right. It's more than that, however. It also means being aware of what's happening to the industry and what your job's going to be in the future. Knowledge encompasses keeping abreast of technical developments, what's possible and what isn't, and changing styles in attitude and approach.

Any time you're hired to script a project, the person who hires you assumes that you have the kind of knowledge it takes to do the job. To that end, you need to soak up every fragment of information you can about the business, whether it concerns writing or not.

I can give you no better example than the rise of interactive. It's created a whole new field.

It's also left many scriptwriters behind. Because they know how to do film or slide shows, they take it for granted that that's enough. *Flow chart* and *branching* are alien concepts.

Then they wonder why, year by year, their pickings get slimmer.

## Reliability

There's a traditional question in the film business: "Do you want it in a hurry or do you want it good?"

Or even more to the point, "Are you reliable? Can we count on you to deliver the script on schedule? Is your word good?"

Your track record tells the story. Has anyone ever had to hold off shooting because you simply didn't complete your script on time? Has inadequate or faked research held up production or embarrassed clients?

That kind of thing is weighed for you or against you. It counts, as when in Mexico a man approached me with a truly glowing proposition. Though years had passed, I remembered how he'd performed—or *not* performed—in other deals where I'd trusted him, and so I turned him down.

Insight, knowledge, reliability. Those are really the things you sell. It's worth your while to remember them.

## DEALS AND HOW TO MAKE THEM

Once upon a time, a novice writer was approached by representatives of a civic club who wanted a slide show on the evils of alcoholism in their community. The writer liked the idea, especially since the price was right. Soon he was up to his eyeballs in books and pamphlets and interviews with psychologists and social workers. His preliminary treatment was received with enthusiasm, acclaim, and loud accolades.

The writer settled down to working up the script, then proudly carried the finished product to his mentors. The reception was again enthusiastic.

The writer glowed. Then, although he hated to introduce such mundane matters, he asked about when he would receive the agreed payment.

His mentors referred him to the sponsoring group, where he learned an interesting fact. Although the project indeed had been discussed, final approval never had been given. The treasurer, a power within the organization, stood firm in his opposition to the project.

To make a long and painful story shorter, the writer never did receive his money. That I can still remember it so clearly after more than forty years tells you how much it galled me.

I hope that you, as writers, never will be faced with a similar disaster. To that end, let's consider a few of the steps you can take to prevent it.

### The Range of Contracts

Contracts—or perhaps I should say agreements—come in a variety of forms. Most writers have done jobs under terms ranging from a producer's or sponsor's suggestion of "Why don't you work that up?" to documents so multiclaused and replete with legal jargon as to make one's hair stand on end.

Each of us has a chosen way of dealing with this issue. My own tends to involve avoiding either extreme in favor of a simple letter of agreement in which I outline what I propose to do, for how much, and under what terms. When my client signs and returns a copy of this to me, acknowledging acceptance, the deal is made. Since each project is different, so is each letter, though the difference is more in details than in outline.

There are two reasons for my preference for such a letter rather than the simple handshake type of agreement. First, your client may die or quit or be transferred. The successor may either be unaware of or choose to deny your project's existence. Your letter protects you from this.

Second, misunderstandings arise in even the most cordial relationships. When that happens, it is nice to have some point of reference to fall back on—not hair-splitting legalese but just general terms to remind each of you of the details that might otherwise be forgotten.

What is wrong with the other end of the stick, the total legal instrument? First, I am not a lawyer, so I really would not know how to write or read such a document. Paying a legal eagle to draw one up would take too large a proportion of the profit out of most script jobs.

Even more to the point, however, is my conviction that no legal contract really means much. That is to say, I enter any agreement with the intention of fulfilling my part of the bargain. I do my clients the courtesy of assuming that such is also their intention. Indeed, if I do not trust them, I prefer not to do business with them at all. Life is too short to waste trying to second-guess thieves.

It would not be true to say that I never deviate from this approach. Often, when old and trusted friends have called, we have never gotten around to putting anything in writing. Again, when I'm dealing with some major government agency or corporation, I recognize that their red tape is sacred to them and so solemnly affix my signature to some ponderous compilation of verbiage which, in the end, means no more than my one- or two-page statement.

## Four Things to Watch

Within your contract, however you draw it, you should cover at least four things: how much you get, how you get it, cutoff points, and approvals. Of course, you should also have a statement as to what you are supposed to do as your part of the bargain.

How much you get paid is a somewhat involved subject that we will take up in the next section. How you get your money is a simpler matter because I can give you some definite advice. Specifically, *accept no assignment that does not pay you as you go along.*

The woods are full of AV projects that never have been completed and never will be. In each case, the writer who agreed to script the job "on speculation"—that is, on the gamble that it would be produced and make money and that he or she thereupon would be paid a set fee, a share of the profits, or both—ended up with empty pockets and a lot of time lost.

I'm not proclaiming this from any superior height. I too have listened to honey-tongued promoters and been persuaded to work up ideas that just could not miss. The only trouble was, they *did* miss—and there went a week or a month or more of work unrewarded, as surely as if that much cash had sifted through a hole in my pocket. Now when tempted to gamble, I try to limit myself to proposals for books, which I can peddle from publisher to publisher until they hit. Where AV ideas are concerned, I'll talk up an idea to a producer for an hour any day but that is as far as I go. Beyond, I want to see money on the line.

But back to our point: *Accept no assignment that does not pay you as you go along.* How do you set this up?

A good system for arranging fees, it seems to me, is to call in your letter of agreement for three equal payments: one for research and development at the start, when the agreement's signed; one for working up the proposal and treatment, payable when you deliver them; and one for preparing the production script, again payable on delivery.

You may split payment into quarters instead of thirds, with final payment on completion of the project, but this can be hazardous, since you have no way of knowing when or if production will be finished.

Payment on delivery of treatment or script does not mean that the client is stuck if your work is unsatisfactory. I take it for granted that, within reason, you will try to straighten out any kinks that develop. If you fail—well, we will consider that when we talk about cutoff points.

In many cases, as a beginner you will not be able to get these terms. You certainly should strive for them, however, and you certainly should not work on pure speculation unless you are at that total novice stage where experience is its own reward for you.

In any event, do not hesitate to point out to your client that research, for example, is tedious and time-consuming work. The proposal and treatment center on creativity, conceptualizing, and talent. Precise description of visuals in the production script, logical development, and smooth handling of narration cut production costs sharply.

Consequently, the payment for your work should not be tied to someone's whim. A script is the bedrock of any project. If the client is serious, he or she should be willing to invest your fee to get it.

A *cutoff point* is, in effect, a contract cancellation clause. It permits a client to end a writer's connection with a project if his or her work proves unsatisfactory.

This is entirely legitimate. The client has a picture in mind of the desired script or, at least, can recognize the undesirable ones. If it becomes obvious that the client and the writer are just not on the same wavelength, therefore, the client can pay for the work done up to that point and cancel the agreement.

If the client is going to cancel, ordinarily he or she does so on completion of the treatment. Cancellation does not necessarily reflect in any way on the writer's competence; it may result from anything from personality conflict to loss of anticipated financing.

Cutoff points should be specified in your letter of agreement, just to avoid possible misunderstandings. All it takes is a simple statement like, "This agreement may be canceled by either party upon written notice following delivery of and payment for treatment." The cutoff point could equally well be at the proposal stage or first-draft screenplay (if you are writing for film) or whatever.

Suppose you and the client get along famously and make it through all cutoff points without event. Your agreement should still include a paragraph calling for approvals—*sign-offs* as they are sometimes termed—on completion of each stage. In other words, when you finish the proposal and the client cries, "Gee, that's great!" you should also receive some sort of written statement that says, "Proposal is satisfactorily completed and approved." A similar statement should also be written for the treatment and shooting script, and maybe even the final narration.

It's quite possible that this sort of formality is not important if you are dealing with a small producer on a one-to-one basis. But it may be. Producers and sponsors alike have been known, in the later phases of a project, suddenly to become wildly enthusiastic about a new idea. Forgetting the fact that the writer already has devoted a considerable number of hours to preparing a first version, they demand that he or she proceed with the new concept as if the original never existed.

This just is not acceptable. So the writer needs to be able to produce a signed approval for each step of the job already completed, backed up by an agreement clause that says any major or nonroutine changes required after approval are to be paid for as if a new project were being launched. This can make the world much brighter for the writer.

## The Right Price

I write this with a simple time sheet beside me. Of loose-leaf lined paper with a place for a project title at the top, it is divided into five columns headed Date, Aspect, From, To, and Time. The column for Aspect is wider than the rest and details the particular phase of the job I am working on: planning, proposal development, treatment development, production script, etc. Sitting down to work, I fill in the date, aspect, and starting time. When I finish, I scribble the quitting time under To and hours worked under Time.

Hardly an elaborate system, it still enables me to know pretty much the number of hours I've devoted to a given endeavor. This is important because, when I divide it into the price I got for the job, it tells me how much per hour I have earned.

This figure is, obviously, not very accurate. You cannot cover the minutes you spent trying to devise an incident or an angle as you drove from here to there, or that sudden wide-awake period in the middle of the night when you scrawled a solution to a problem on a scratch pad. The figure is, however, better than picking a price out of thin air—that's the way to disaster. (I wrote a script some years ago, only to find that in the end I'd netted—so help me!—a grand total of 43 cents per hour on it.)

I have two other reasons for favoring this sort of time-analysis approach. First, having at least a rough idea of the hours involved in carrying through a particular type of assignment—be it slide shows or suspense novels—enables you to allocate your time more accurately and bid more intelligently next time that sort of chore comes your way.

Second, too low a bid for a job merely convinces your client that you do not think you are worth much, and that does neither of you any good.

In calculating your bid, how much you should charge per hour is, obviously, a matter for personal decision. An approach that makes sense is to base your figure on the amount you could earn at a regular job. Thus, if you have reason to believe an employment agency might place you at fifteen or twenty dollars an hour, then surely your work as a writer should rate at least a similar amount.

When freelancing full time, another factor needs to be considered: There will be times when you have no assignments and yet you'll still have to eat and pay rent. Your script fees, therefore, must cover this dead time as well as days you actually work. Hours spent soliciting assignments must also be entered, as must office and equipment overheads and the like. All should be adjusted to take account of the going rate, if any, in your area.

None of the above is designed to lay out ironclad rules for you. My only goal is to nudge you in the direction of treating AV writing as a business. Believe me, you need to do so if you are to survive!

Two final items should be mentioned in regard to this matter of price and pricing: travel and cutoffs.

Travel, particularly, is an item to consider in every bid. It not only increases your costs when you have to work away from home, it may also knock you out of the running in competitive situations. I have often been able to snag a local job just because I could bid $500 less for not having to add in airfare, car rental, and hotel bills.

Even more important, writing away from your home base cuts you loose from contacts and research materials while, on the other side of the coin, attempting to carry out an assignment by remote control invariably involves you in situations in which you have to make extra trips to confer with specialists or technical advisors, resolve questions of authority, or time out tricky sequences.

This is not to say you should not travel, but you do need to include it in your budget.

The cutoff is, as mentioned earlier, a perfectly legitimate contract provision. Occasionally, however, you will encounter a sponsor or producer who will use it in a less than forthright fashion. For example, you agree to prepare a presentation. When you submit your treatment, however, the sponsor rejects it on whatever grounds. Later you discover that the sponsor is not only perfectly happy with it but that he or she proposes to shoot the program from it.

Whether you are paid for the treatment or not, you are entitled to be irked at this procedure on two counts: artistic and financial.

Artistically, the issue is that the client probably will not do much of a job when it comes to expanding the treatment into shots. The fact that he or she will go to such extremes to dodge your fee is hardly indicative of orientation to a quality product. If you are like me, you like anything you do to stand a fighting chance of looking good.

Financially, you are being treated very shabbily indeed. The preparation of the proposal and treatment constitute the most creative—and nail-gnawing—aspects of scripting any program. Conceiving an approach and thinking through its development are the really difficult parts of audiovisual writing.

Further, by the time you have written the treatment, a high proportion of the shots for the final script already are blocked out in your mind. The fact that they are not yet down on paper is primarily a matter of typing. To be knocked out of this portion of your fee on meretricious grounds falls just short of highway robbery.

How should you deal with situations like this? As outlined above in my comments on contracts, *insist* on an arrangement that gives you one-third payment on signing, for research and development of the project; a second third on delivery (*not* approval) of treatment; and the final third on submission of the shooting script. By handling the deal in that way, you are assured of at least two-thirds of your full fee, even if you are cut off after the payment for the treatment.

Incidentally, all this is not to imply that most sponsors or producers will resort to such attempts to bilk you. On the contrary, the overwhelming majority are totally honest. As in any business, however, there always are a few who cut corners and these are the ones you need to know about when you are beginning.

Entirely apart from chicanery, doing business as a writer has other hazards of which you should be aware.

## Danger! Specifications Ahead!

They stick the strangest clauses into AV script contracts sometimes. As a writer, you need to be acutely aware of all of them. Nowhere is the old line about "Read it before you sign it" more vital, because you will have to live with those terms. Failure to understand or pay attention to them can commit you to totally impossible conditions—and that is a conservative statement! Overlooking an apparently routine detail like "The script will include frame counts that must coincide with the timing of the narrative" may prove disastrous.

In general, specifications fall into two categories: *technical* and *image*.

Technical specifications are pretty much the kind of thing you would expect: lip-sync dialogue, voice-over narration, or a combination; age level for narration vocabulary; music included or excluded; and visuals of particular equipment or action.

(Visuals of equipment can be a headache. You need to find out exactly where the equipment may be observed in operation. Down the street is one thing but if it can be seen only in London, Montreal, Manila, or Lower Slobbovia, some travel money needs to be written in. Similarly, are you to center on representative plants or parks or state prisons, or must all types be pictured? If all, how much attention must be devoted to each? Will single slides be adequate, or is glowing descriptive narrative for individual operations required?)

It is in image specs that contracts can really go crazy. My wife worked on a show that specified that all the physicians were to be American Indians. When nurses and other health professionals were shown, Indians were to be in the majority. Caucasians were permitted only in incidental roles.

You will also encounter clauses that require women in certain parts; that specify a certain percentage of Blacks; that restrict dialogue forms of address to Mr. and Ms.; or that exclude drinking, smoking, profanity, and fur coats. You name it, it has appeared in a writer's contract somewhere.

Be aware of these restrictions and their possibilities for handicapping you.

## The Matter of Schedules

The same principle applies to time schedules. Study them carefully and be sure you can meet them.

How can you be sure? Partly, you judge by the material. Are you sufficiently familiar with the program's subject that you can run it through in a hurry or is it something that will involve you in endless reading, field work, and conferences with experts?

Equally or more important are the time sheets I described earlier in this chapter. Nothing can help you more in appraising a time schedule than your knowledge—based on past performance—that you can come up with a generally acceptable proposal of the type required in three days, expand it into a treatment in five more, and then produce the final script at the rate of ten pages per day.

Producers like it when you can talk in such terms. They see it as the mark of the professional. The writer who is unsure of these matters, on the other hand, they see as the eternal amateur—a genus they tend to view as unreliable and walk wide around.

It goes without saying, of course, that no estimate you make should pretend to be absolute. As a rule of thumb passed on to me years ago says, "Always write bad luck into your contract."

My prize example of this centers on an incident that occurred while I was living in Mexico. A firm with which I had had happy relations for years sent me an assignment for a series of AV programs to be prepared from tape recordings. Since it was a time when cash was in short supply for me, I was delighted. I cleaned up my typewriter and prepared to settle down to work.

There was a catch, however. The tapes did not come—and the job had a rigid deadline.

Finally, I got on the telephone to find out what had happened. No one could give me any answers. Even worse, no transcript of the tapes was available.

Days passed, then weeks, then the deadline, and still no tapes appeared. Months later, they did arrive. They had, it turned out, been held by Mexican Customs as dutiable items. I had to pay a ridiculous fee to get them out of the post office.

By then, of course, it was too late. To make matters worse, I lost not just the assignment but my edge with the company. As they explained, if I chose to live in a place where mail did not even reach me on schedule, they could see no way to utilize my services further.

A second case in point. Today a favorite client told me that henceforth all productions would be done in-house, so there'd be no further need for my services. The only exceptions would be if we could produce three-dimensional graphics.

To do that I'd have to have specialized computer equipment. *Expensive* equipment. It put me out of the market.

Such things happen all the time in the AV field. More and more frequently a freelancer is faced with the necessity of having equipment that's compatible with the system of the production company for which he or she is working.

Yet even if you buy an expensive system with specialized capabilities, it may not prove compatible with that of the next client or it might quickly be rendered obsolete by new developments.

The answer? Establish a relationship with a group of clients and buy equipment compatible with their equipment. Even better, arrange to use the company's equipment when you hire on for a project. That avoids the purchase of costly items that you can't use for another job or that fall prey to obsolescence.

Another answer is to stick with simple scripts. Limit yourself to projects you can handle with inexpensive systems. Lower overhead balanced against lower income sometimes equals higher profit for the freelance writer.

Snowstorms, sickness (your own and others'), equipment problems, and unavailability of people or information all go to emphasize that you should "Always write bad luck into your contract."

# Afterword

It is, of course, the best job in the world.

And the worst.

The easiest.

And the hardest.

The most satisfying.

And the most frustrating.

Naturally, what I'm talking about is audiovisual scripting. But not *just* scripting, for what I say applies equally to almost any creative activity.

Most work today involves your time, your muscle, and maybe even your brain. AV scripting takes vision also.

Do I need to tell you that this moves AV scripting over to the unique side of life's ledger? Vision is neither common nor does it always come easily. It brings with it the gnawing anguish of ideas that refuse to surface when you need them, and ideas that do surface but then are rejected by your colleagues or clients.

That kind of thing can be hard to take. Many find it too painful to endure.

On the other hand, if you spark to the excitement that ideas bring, if you enjoy the stimulus of vision and of seeing things not as they are but as they might be, if you thrill to the challenge of finding new ways to catch interest and reach viewers, then maybe scripting *is* the kind of calling for which you are cut out.

To put it another way, if what you want is work without strains and stresses, something you can leave behind when you walk out of the shop, scripting is not for you. Whereas, if on occasion you lose track of time because your work is your fun also, if "mounting, goal-oriented inner tension" sometimes counts for more with you than mindlessness or a cushy spot, then consider scripting.

Scripting is a field in which your success, your failure, your pleasure, and your fulfillment *all* depend on you. Nowhere are you more in command of your own destiny.

I hope you like that idea and you have the intelligence and imagination and backbone to stick with it. This is a weird and wonderful game you are sitting in on—poker with everything wild. The sky's no limit.

# Glossary of AV Idiom

No one's going to be totally happy with this glossary. I've put in too many things you'll never encounter—and I've left out too many that will assail you on every side.

This is simply another way of saying that this can be a complicated field. This glossary would be too long if everything were covered, especially since the technical and scriptwriting fields so often overlap. It therefore is limited to things a scriptwriter needs to know or will most often come across. The goal is to give you at least a portion of the language you may encounter in talking about the field.

Will you need all these terms in order to write scripts? No. But it *will* help if you have enough command of the trade jargon to be able to understand what's going on and to be able to talk intelligently with your colleagues.

**Adaptation**　Presentation in one medium of work originally designed for another, as in a film developed from a novel.

**Analog signal**　A continuously variable signal in a wave pattern. The amplitude and frequency of the signal vary with the amplitude and frequency of the sound wave, for example. This is the most common signal today; it is used in televisions, radios, telephones, video cameras, and microphones, among other devices. Noise is a problem with an analog signal.

**Angle**　The positioning of a camera in relation to the subject: high angle, low angle, 45° angle, etc.

**Animation**　The simulation of life or movement in inanimate objects on film, video, computer, or other means of recording.

**Answer print**　The first combined picture and sound print, in release form, of a finished film.

**Art director**　The person who designs and supervises set construction for a film or linear video.

**Audio**   The sound portion of an AV program. It includes narration, music, sound effects, and dialogue.

**Audioconference**   A form of teleconference where all participants can hear each other, but not all can see each other.

**Audiovisual**   A program that includes both sound and pictures.

**Authoring**   Programming for computer-based training or interactive video.

**Authoring software**   A specialized computer program that enables non-programmers to use a computer to develop an interactive computer-based or videodisc program.

**Background (BG)**   Sound, setting, or action subordinate to a given scene's foreground or dominant elements.

**Bells and whistles**   Flashy elements in an interactive program.

**Branching**   (1) In an interactive program, offering two or more possible courses of action between which the viewer must choose to move ahead in the program. (2) Moving from one sequence in an interactive program to another.

**Business**   Action introduced to build up or reinforce characterization, a sequence, a plot point, or the like.

**Busy**   Refers to a setting or an action that distracts from the desired impression through the inclusion of unnecessary detail.

**CAI**   Computer-aided instruction. *See* Computer-based training.

**CARDS**   Computer-Assisted Repair and Diagnostic System.

**CBT**   Computer-based training.

**CD**   Compact disc.

**CDI**   Compact disc interactive.

**CD-ROM**   Compact disc, read-only memory.

**CDV**   Compact disc video.

**Closeup (CU)**   A picture that emphasizes a particular feature of a subject by showing it in disproportionately large size.

**Commentary**   The voice-over spoken remarks or explanations that often accompany a program's visual presentation.

**Compact disc (CD)**   A digital audio disc used primarily for music. It cannot be edited.

**Compact-disc interactive (CDI)**   A digital optical disc that can store both video and audio signals and can play back recorded information in segments in response to a user's instructions, rather than in a predetermined order. It cannot be edited.

**Compact disc, read-only memory (CD-ROM)**   A digital optical disc used to store and play back information. It cannot be edited.

**Compact disc video (CDV)**   A digital optical disc that can store and play back twenty minutes of CD-quality music and five minutes of video.

**Compressed audio**   Audio that is digitally recorded on individual disc frames, making more audio available per disc.

**Computer**   A data processor that carries out preplanned data manipulations under the control of a program.

**Computer-based training (CBT)**   An interactive lesson prepared to be delivered by a computer.

**Computer conference**   A meeting where participants are linked via computers. They may send documents or information to one another, and they may communicate by keying information into their computer keyboards.

**Continuity**   (1) The sequence of events to be presented in a program. (2) The smooth linking of one event or picture to another.

**Credits**   Titles that name the people who worked on a program.

**Cue**   A signal to begin some specified fragment of a program—a movement, a speech, a strain of music, or the like.

**Cut**   Instantaneous change from one shot to another.

**Cutaway**   In continuity cutting, a shot that does not include any part of the preceding shot, as when a character glances out a window and the shot that follows shows what he sees.

**Cutback**   The shot that follows a cutaway if it returns to the preceding action, as when a cutaway shot of what a character sees out a window is followed by a shot that includes the character as he turns away from the window.

**Database**   The collection of information that can be accessed by a computer.

**Digital signal**   A series of on/off pulses that can be coded to carry both video and audio information. A digital signal can be processed by a computer, manipulated by a video camera, and stored on floppy disks, tapes, and optical discs.

**Digital video interactive (DVI)**   A digital optical system that can also store material on a computer's hard disc drive.

**Director**   The person who supervises a program's production, translating the script into audiovisual form.

**Disc (disk)**   A circular piece of plastic used to store computer information. *See also* Compact disc; Compact disc interactive; Compact disc, read-only memory; Compact disc video; Videodisc.

**Dissolve**   An optical effect in which a fade-in is superimposed over a fade-out so that one shot replaces the other.

**DVI**   Digital video interactive.

**Electronic publishing**   The preparation of material to be delivered to its audience via electronic means. Material is considered published if it is available by modem or on computer discs or such.

**Establishing shot**   A shot that makes clear the relationship of one element in a picture to another. Ordinarily, this means that it takes in all or a considerable portion of the setting in a long shot.

**Fade**   An optical effect used as a transitional device, in which the picture on the screen gradually goes to black (fade out) or takes form from black (fade in).

**Film**   A strip of celluloid that is used to record pictures.

**Filmstrip**   Still photographs or graphics presented on a strip of 35mm film, with or without accompanying sound.

**Floppy disk**   A circular, magnetic platter used for recording computer-generated data.

**Flow chart**   A diagram tracing the movement through a presentation. These are commonly used for interactive preproduction planning.

**Foreground**   Action, objects, or settings closest to the camera in a given shot or shots.

**Format**   An established pattern for presentation of a particular type of script or AV program.

**Frame**   (1) One of the individual still photos that, collectively, constitute a motion picture. (2) To compose a shot to include, exclude, or emphasize certain things, as in "This shot is framed so that the narrowing distance between the two riders is immediately obvious."

**Freeze-frame**   A single frame from a motion sequence that is held on the screen, thereby stopping the action.

**Graphics**   Visuals, drawings, and text used in a production.

**Hardware**   The physical equipment used to process data or display a videodisc, videotape, or computer presentation (i.e., the machinery).

**Hook**   (1) A striking incident, unique action, or the like, which is used to capture audience attention at the beginning of a program. (2) A linking device (question–answer, repetition, agreement–disagreement, etc.) used to render dialogue cohesive by tying each speech to the one ahead of it.

**Icon**   A symbol that represents a function, usually in videodisc or computer technology.

**Imax**   A special 70mm high-fidelity motion-picture system that creates gigantic images, supplemented by a six-channel stereo sound system.

**IN, UP, DOWN, OUT, etc.**   Terms used to describe the introduction, conclusion, and handling of the volume of a program's sound elements.

**In-house**   Refers to work done within a company using its own employees.

**Insert**   A shot, frequently an extreme closeup, of some object or detail (a letter, a picture, a hand, a tattoo) that can be filmed in such a manner as to eliminate the setting. Thus, it may easily be shot out of sequence and cut into the picture during editing.

**Instructional design**   The process of determining the teaching methodology to be used for the presentation of material in an AV program.

**Instructional designer**   The person in charge of an interactive (or other teaching) program.

**Kiosk**   A self-contained video unit that is user-operated or operated automatically.

**Laser disc**   Generic name for a reflective optical videodisc format.

**Layout**   The visual presentation of a proposed display used for planning.

**Lead-in**   A program's visual or audio introduction to a particular subject or aspect of a subject.

**Linear**   A video or audio sequence designed to be played from beginning to end without branching or changing.

**Location**   A natural setting (as contrasted with a sound stage) where a program, in whole or in part, is to be shot.

**Long shot**   A picture that relates a subject to its background, as in an establishing shot.

**Loop**   (1) Film or tape packaged in a cartridge for special projects, and spliced end to beginning to permit continuous repetition. (2) A program on a video or computer disc that is programmed to repeat.

**Mastering**   The combination of a program's sound and picture in final form, from which copies are made.

**Medium shot**   A picture dominated by a particular subject (person, house, car, etc.) with only incidental background.

**Menu**   A table of contents of a computer program or video disc.

**Monitor**   A screen that will accept video signals, computer display information, or both.

**Montage**   (1) An assortment of photos, artwork, or stills arranged on a display surface for visual interest. (2) In film or video, fast cuts and optical effects combined to build emotion or provide time or space transitions.

**Multi-image**   An audiovisual program that projects more than one image at the same time.

**Multimedia**   An audiovisual program that uses a variety of media as sources, such as film, animation, video, and slides.

**Multimedia program**   One that combines a variety of audio and video techniques (frequently including split-screen work, multiple soundtracks, etc.) in a pattern designed to seize audience attention.

**Multivision**   British for *multi-image*.

**Narration**   The commentary that the narrator delivers in a voice-over.

**Narration script**   Commentary prepared for a narrator in a voice-over recording situation.

**Narrator**   The person who delivers the voice-over comments that accompany a program's visual presentation.

**Off screen (OS)**   Action or sound closely related to a picture but not included in it, as when an unseen character cries, "Help!"

**On-line**   Equipment in direct communication with a computer.

**Optical disc**   A videodisc that uses a laser light beam to read information from the surface of the disc.

**Optical effect**   A change in a film's pictorial image (fades, dissolves, wipes, etc.) that is created in the laboratory or by computer, rather than in the camera or editing room.

**Orientation shot**   Another name for an *establishing shot*.

**Overlay**   (1) To superimpose text or graphics onto motion or still video. (2) An acetate cover sheet that adds words or pictorial elements to an existing picture.

**Pay off**   To make significant use of something previously planted, as when at the climax of a film the protagonist faces down the villain with a previously planted pistol.

**Performance-based training**   An interactive program that will not advance to the next sequence until the user has attained a certain skill level in response to program prompts.

**Performance certification**   Interactive training that automatically records and makes available a record of the user's correct and incorrect responses for each attempt to complete a segment or program.

**Pixel**   Picture element. The smallest component of a video image that can be manipulated by a computer.

**Plant**   An apparently offhand establishment of an idea, character, or property to be used more significantly later in the program—a gun seen in a drawer when the drawer is opened to get a pad or pencil, for example.

**PMMA**   Polymethyl methacrylate. The plastic used to manufacture many videodiscs.

**Point of purchase (POP)**   Interactive video kiosks set up in commercial locations where customers can actually buy the products shown on the kiosks.

**Point of view (POV)**   A picture from the position of a particular character.

**Postproduction**   The work done after the actual filming, shooting, programming, and other preparation has been completed on a video or film presentation. It includes editing, encoding, and special effects.

**Premise**   A hypothetical "What if . . . ?" that provides the basic idea for an AV production.

**Preproduction**   All design and planning stages before the actual shooting of an AV presentation. It includes the creation of flow charts, storyboards, and scripts.

**Producer**   The person who supervises and coordinates a program's overall production.

**Production**   The actual filming, shooting, programming, and other preparation of a video presentation.

**Production script**   *See* Shooting script.

**Program**   (1) To write a series of codes and instructions into a computer so that it will respond to instructions in a preplanned manner. (2) The set of prepared instructions to be used in a computer for predetermined purposes. (3) A show or presentation previously prepared for viewing by an audience.

**Programmed instruction**   Instruction designed to begin with simple concepts and to advance step by step in a logical progression to more complex concepts.

**Programmer**   A person who prepares a program for a computer.

**Proposal**   A summary of a proposed AV project, designed to be as interesting and persuasive as possible.

**RAM**   Random-access memory.

**Random-access memory (RAM)**   The part of a computer's memory in which is stored information that can be accessed, displayed, and edited or changed.

**Read**   To locate and display information from a tape or disc to a screen or monitor so it can be used by the viewer.

**Read-only memory (ROM)**   (1) The part of a computer's memory in which essential operating information is stored and can be read, but not altered. (2) A means of storing information in a computer's memory so that it can be accessed and read but not written to or changed.

**Real estate**   The space on a videodisc or tape.

**Real time**   The chronological time required for a program to run.

**ROM**   Read-only memory.

**Scriptwriter**   A person who writes the instructions for the production team in an AV presentation.

**Search**   To look for a specific frame on a disc or tape. This is a part of interactive programs where branching occurs.

**Segment**   In an interactive program, material meant to be accessed together to form a unit of information, or the information between a start and a stop frame.

**Sequence**   (1) A related series of shots, unified by some element they hold in common—setting, concept, action, character, or mood, for example. (2) A major segment of a film or linear video, on the order of a chapter in a novel.

**Sequence outline**   A list and brief description of the sequences to be included in a film or video.

**Shooting script**   A script that describes each picture to be shot for a given AV presentation. It serves as a guide for production. Also called a *production script* or *shot list*.

**Shot**   A single picture. In film or video, those pictures recorded by a motion camera in a single run.

**Shot list**   *See* Shooting script.

**Simulation**   A representation of a real activity or event. Simulations are used to let users observe or participate in activities that might be dangerous, difficult, or very expensive for them to do in real life.

**Slide**   One frame of 35mm film, developed and mounted for display.

**Slide show**   A series of 35mm slides, arranged in a predetermined order and (ordinarily) presented with interpretive voice-over commentary.

**Software**   Any programs or procedures used in the operation of a computer—as opposed to hardware, which is the equipment itself.

**Sound effects (FX, SFX)**   Any sound other than voice or music (wind howling, engines roaring, or a knock at the door) included in a program.

**Special effects**   Photographic effects obtainable only under controlled conditions and through the use of specialized techniques. Many are computer-generated today.

**Split screen**   Different pictures shown on portions of a single screen at the same time.

**Sponsor**   The person, group, or organization for whom an AV program is produced; generally, whoever puts up the money.

**Still frame**   Information recorded on a single frame of a videodisc that is intended to be displayed as a single motionless image.

**Sting**   A sudden, emphatic sound, often musical, used to punctuate an AV presentation.

**Stock shot**   A picture taken from existing sources rather than shot specifically for a given program.

**Storyboard**   A plan or outline of a program set forth in sketches or still photos.

**Subject matter expert**   An individual with detailed knowledge of the subject of an AV presentation, who determines what information will be included in consultation with the instructional designer and the scriptwriter.

**Submenu**   A menu within a program that refers to only one portion or segment of the entire program. It is frequently used to let a viewer review a segment or to branch within one segment without returning to the main menu for the entire program.

**Superimposition**   One photographic image placed on top of another to create a new effect.

**Synchronization**   The precise matching of sound to action, as when the voice fits the lip movements exactly.

**Synopsis**   A brief outline of a proposed AV program's content.

**Target audience**   The group to which an AV or other presentation is intended to appeal.

**Telecommunication**   Usually refers to communication between two computers via telephone lines.

**Teleconference**   Any meeting where participants are separated by distance and linked by electronic means.

**Titles**   Screened words that provide information about a program.

**Touch screen**   A video or computer screen that responds to signals when touched in specific areas and directs changes in the program being viewed. Touches may signal branching in interactive programs, for example, or they may merely cause the program to stop or to continue.

**Treatment**   Summary of a proposed project's contents, elaborating on the project proposal and "selling" it to the client.

**Tutorial**   A program designed to be used by a single individual for training purposes.

**Two-shot (three-shot, etc.)**   A shot that includes two (three, etc.) people, often in conversation.

**User-friendly**   (1) Computer programs that are easily understood by a person with little computer experience. (2) Any interactive program with operating instructions that are easily understood and followed by a novice in the field.

**VCR**   Video cassette recorder. The home device used to play and record videocassettes.

**VHS**  Video Home System. A videotape format developed by JVC for noncommercial use.

**Video**  The visual aspect of an AV program. A system of recording and transmitting information and pictures by translating moving or still images into electrical signals. Although technically the term refers only to the handling of pictures, it is now commonly used to refer to entire programs recorded electronically, including sound.

**Videocassette**  A case holding magnetic tape for recording, playing, or storage, usually for noncommercial purposes.

**Video computer-based training**  Computer-based training that includes graphics or video frames.

**Video conference**  A meeting of participants in different locations where the participants are linked by an electronic means that permits visual exchange. One-way video conferences are conferences where video of only one segment of the participants will be sent to all participants. Usually, telephone lines are used for two-way audio communication at one-way video conferences. Two-way video conferences are where all participants can see and hear each other during the conference.

**Videodisc**  A thin, circular plate on which video, audio, and data signals can be recorded for playback or storage.

**Videotape**  Magnetic tape that records both sound and pictures.

**Video wall**  A picture display using an electronic "splitter" to split a single video signal across the required number of monitors.

**Visual aids**  Any object that is seen and that accompanies an audio presentation, including slides, overhead transparencies, and video images.

**Voice-over (VO)**  Spoken lines from an unseen source that accompany a program's visuals.

**Wild shot**  A video shot or an audio segment recorded independently of each other, with no attempt to synchronize.

**Window**  (1) A portion of a screen where a video image may be displayed independently of the principal image. (2) An entry segment of an interactive videodisc program.

**Write**  To transcribe recorded data from one place to another. Most frequently used to describe movement from computers to discs or among discs.

**Write once/read many (WORM) optical disc**  A digital optical disc used to store computer-compatible information. It allows a user to record data on the disc once, but does not allow that data to be edited or changed after recording. On the other hand, the material can be accessed many times.

**Zoom**  To move in to a subject so that its relative size on the screen increases; or to pull away from a subject so that its relative size on the screen decreases.

# For Further Reading

Bruner, J.S. *Towards a Theory of Instruction*. Cambridge, MA: Harvard University Press, 1967.

Electrosonic AV, Ltd. *What Is Audio Visual?* London: Electrosonic, 1983.

Fielding, Raymond. *The Technique of Special Effects Cinematography*, Fourth Edition. Boston: Focal Press, 1985.

Fleming, M.L., and Levie, W.H. *Instructional Message Design: Principles from the Behavioral Sciences*. Englewood Cliffs, NJ: Educational Technology Publications, 1978.

Guthrie, E.R. *The Psychology of Learning*. New York: Harper & Brothers, 1952.

Halas, John (ed.). *Visual Scripting*. Boston: Focal Press, 1976.

Knapper, Christopher K. (ed.) *Expanding Learning Through New Communications Technologies*. San Francisco: Jossey Bass, 1982.

Knirk, Frederick G., and Gustafson, Kent L. *Instructional Technology: A Systematic Approach to Education*. New York: Holt, Rinehart & Winston, 1986.

Madsen, Roy. *Animated Film*. New York: Interland Publishing Inc., 1969.

Mezey, Phiz. *Multi-Image Design and Production*. Boston: Focal Press, 1988.

Mirabito, M., and Morgenstern, B. *The New Communications Technologies*. Boston: Focal Press, 1990.

Nievergelt, Jay, et al. *Interactive Computer Programs for Education: Philosophy, Techniques, and Examples*. Reading, MA: Addison-Wesley, 1986.

Sleeman, Phillip J., et al. *Instructional Media and Technology*. New York: Longman, 1979.

Soulier, J. Steven. *The Design and Development of Computer Based Instruction*. Boston: Allyn & Bacon, 1988.

Swain, Dwight V. *Techniques of the Selling Writer*. Norman, OK: University of Oklahoma Press, 1974.

————, with Swain, Joye R. *Film Scriptwriting: A Practical Manual*, Second Edition. Boston: Focal Press, 1988.

Taylor, John, and Walford, Rex. *Learning and the Simulation Game.* Beverly Hills, CA: Sage Publications, 1978.

Utz, Peter. *Today's Video: Equipment, Setup, and Production.* Englewood Cliffs, NJ: Prentice-Hall, 1987.

The Videodisc Monitor. *Videodisc and Related Technologies: A Glossary of Terms.* Falls Church, VA: Future Systems, Inc., 1986.

# Appendix:
# Useful Addresses

Here are some of the organizations, institutions, publications, and the like that you as an AV scriptwriter might want to contact. This list is of course incomplete, but at least it's a start in your continuing education.

AV Video, 25550 Hawthorne Blvd., Suite 314, Torrance, CA 90505.

Corporate Video Decisions, PO Box 6018, Duluth, MN 55806-6016.

Eastman Kodak Co., 343 State St., Rochester, NY 14650.

Film and Video Production, 8170 Beverly Blvd., Suite 208, Los Angeles, CA 90048.

Hope Reports, Inc., 1600 Lyell Ave., Rochester, NY 14606.

International Interactive Communications Society, c/o Level 4 Communications, 3 Dallas Communications Complex, Suite N-2, 6311 North O'Connor Rd., LB-134, Irving, TX 75039.

International Television Assn., 6311 North O'Connor Rd., LN-51, Irving, TX 75039.

Knowledge Industry Publications, Inc., 701 Westchester Ave., White Plains, NY 10604.

Nebraska Videodisc Design/Production Group, Univ. of Nebraska–Lincoln, 1800 North 33rd St., Lincoln, NE 68583.

Society for Applied Learning Technology, 50 Culpepper St., Warrendown, VA 22186.

Society of Motion Picture and Television Engineers, 595 West Hartsdale Ave., White Plains, NY 10607.

Sony Video Workshops, Sony Institute of Applied Video Technology, 2021 North Western Ave., P.O. Box 29906, Los Angeles, CA 90029.

Technological Horizons in Education Journal, 2626 South Pullman, Santa Ana, CA 92705.

3M Optical Recording, Bldg. 223-5S, 3M Center, St. Paul, MN 55144-1000.

University Film and Video Assn., address changed annually.
Videodisc Monitor, P.O. Box 26, Falls Church, VA 22046.
Videography, 2 Park Ave., Suite 1820, New York, NY 10016.

# Index